Ethics in Pharmacy Practice: A Practical Guide

Dennis M. Sullivan • Douglas C. Anderson
Justin W. Cole

Ethics in Pharmacy Practice: A Practical Guide

Dennis M. Sullivan
School of Pharmacy
Cedarville University
Cedarville, OH, USA

Douglas C. Anderson
School of Pharmacy
Cedarville University
Cedarville, OH, USA

Justin W. Cole
School of Pharmacy
Cedarville University
Cedarville, OH, USA

ISBN 978-3-030-72168-8   ISBN 978-3-030-72169-5   (eBook)
https://doi.org/10.1007/978-3-030-72169-5

© Springer Nature Switzerland AG 2021

This work is subject to copyright. All rights are reserved by the Publisher, whether the whole or part of the material is concerned, specifically the rights of translation, reprinting, reuse of illustrations, recitation, broadcasting, reproduction on microfilms or in any other physical way, and transmission or information storage and retrieval, electronic adaptation, computer software, or by similar or dissimilar methodology now known or hereafter developed.

The use of general descriptive names, registered names, trademarks, service marks, etc. in this publication does not imply, even in the absence of a specific statement, that such names are exempt from the relevant protective laws and regulations and therefore free for general use.

The publisher, the authors, and the editors are safe to assume that the advice and information in this book are believed to be true and accurate at the date of publication. Neither the publisher nor the authors or the editors give a warranty, expressed or implied, with respect to the material contained herein or for any errors or omissions that may have been made. The publisher remains neutral with regard to jurisdictional claims in published maps and institutional affiliations.

This Springer imprint is published by the registered company Springer Nature Switzerland AG
The registered company address is: Gewerbestrasse 11, 6330 Cham, Switzerland

Foreword

Modern healthcare is becoming more and more complex, as well as the ethical dilemmas that accompany it. The correct handling of these moral problems is more than just a matter of personal opinion, for there are clear ethical principles that inform each of these scenarios. Nonetheless, ethics in pharmacy practice is a relatively new discipline, and only a few textbooks address it.

Traditionally, pharmacists have long been recognized as experts in healthcare law; however this expertise has now expanded to include healthcare ethics. While ethics and law complement each other, they are distinct fields. Ethics is more important than ever in pharmacy practice, for two principal reasons: (1) pharmacy practice standards continue to grow and expand, and (2) the pharmacy practice environment is dynamic, with the pharmacist playing a more central role as a key member of the healthcare team.

In this book, Sullivan and colleagues offer a unique approach to ethical decision-making for practicing pharmacists and students. Unlike other texts, the authors provide clear guidance based on the fundamental principles of moral philosophy, explaining them in simple language and illustrating them with abundant clinical examples and case studies.

The authors of this text are well qualified to address these topics. The principle author is a physician, originally board-certified in general surgery, with advanced graduate training in bioethics and moral philosophy. Both associate authors are pharmacists with extensive clinical and academic experience. All three bring abundant insights from clinical practice, and from teaching clinical ethics and law in schools of pharmacy.

Ethics in Pharmacy Practice: A Practical Guide is divided into two parts: Foundations (Chaps. 1, 2, 3, 4, 5, 6 and 7) and Issues and Cases (Chaps. 8, 9, 10, 11, 12, 13 and 14). The first part presents general ethical concepts and theories of human value, as well as an essential historical survey from Hippocrates to the modern era. It concludes with an analysis of pharmacy professionalism.

The book's second part addresses the major ethical concerns in modern clinical practice, including beginning and end of life issues, rights of conscience, vaccines, and unproven treatments. Chapter 12 includes a "Pandemic Update," with important ethical insights from the 2020–2021 worldwide coronavirus crisis. Chapter 13 features 18 clinical cases and 7 policy scenarios to challenge the reader, while Chapter 14 takes a look to the future, examining genetic ethics and other cutting-edge issues.

Overall, this text places an emphasis on normative ethics and critical thinking as a method for approaching ethical problems. The moral judgment of the pharmacist is essential to this process. The goal is to teach the reader *how* to think, based on ethical principles, not necessarily *what* to think. The case study format provides pharmacists with practical cases and assists in emphasizing the concepts presented throughout the chapters. Educationally, the review and discussion provides topics for instructors to incorporate into their lesson plans.

Ethics in Pharmacy Practice: A Practical Guide may be used as a principle classroom textbook for a graduate/professional level course in pharmacy ethics. However, this text should also be useful to practicing pharmacists and other healthcare professionals who need an additional background in ethics.

The purpose of this book is to help pharmacy students, pharmacists, and other clinical professionals to gain moral competence, professionalism, and compassion. To this end, *Ethics in Pharmacy Practice: A Practical Guide* has lived up to its goal!

Director of Science and Research for the Ohio Jon E. Sprague
Attorney General, BCI Eminent Scholar
of the Center for the Future of Forensic Sciences
Bowling Green State University,
Bowling Green, OH, USA
March, 2021

Preface

Along with physicians, pharmacists are rated highly in terms of public trust and confidence, and in a recent Gallup survey, ranked near the top in honesty and personal integrity (Reinhart 2021). Community-based pharmacists are more accessible than most other healthcare professionals, and there is also a growing awareness that pharmacists add significantly to the effectiveness of primary care initiatives. The impact of the pharmacist on patient health and quality of life has been well-documented. This influence has been undergirded by a professional oath and a commitment to ethical practice.

The focus of the pharmacy profession has changed dramatically during the past half-century. One major change has been the shift from a product-distribution emphasis to a more clinical, patient-centered profession. With the rise of large chain pharmacies, another development has been the transition from pharmacists as mostly self-employed to being almost entirely company employees. The recent growth of collaborative practice agreements has allowed physicians and pharmacists to share responsibilities and treatment authority for many conditions (Advancing Team-Based Care 2021). The increased focus of pharmacists on patient-centered practices will also increase their encounters with ethical dilemmas. This will require not only clinical competency, but also enhanced ethical literacy and a renewed commitment to ethical practice among pharmacists, hence the need for this text.

This book offers a unique and accessible approach to ethical decision-making for practicing pharmacists and student pharmacists. Unlike other textbooks, it gives clear guidance based on the fundamental principles of moral philosophy, anchoring them in their historical context. This text explains these concepts in simple language and illustrates them with abundant clinical examples and case studies.

The strength of this text is its emphasis on normative ethics and critical thinking, that there is truly a best answer in the vast majority of cases, no matter how complex. The authors place high confidence in a pharmacist's moral judgment. The goal is to teach the reader *how* to think, based on ethical principles, not necessarily *what* to think. This means navigating between the two extremes of overly theoretical and excessively prescriptive.

The lead author of this text is Dennis M. Sullivan, MD, MA (Ethics). He is professor emeritus of pharmacy practice at Cedarville University, Cedarville, Ohio. He received his MD from Case Western Reserve School of Medicine in 1978, after which he trained in general surgery, and then served for 2 years of active duty in the U.S. Army at Fort Campbell, Kentucky.

After his military service, Dr. Sullivan worked for 12 years as a medical missionary, providing surgical care in Haiti and in the Central African Republic. From 1996 to 2019, he was an active member of the faculty at Cedarville University. In 2004, he received his MA in bioethics from Trinity University in Chicago. From 2006 until his retirement in 2019, Dr. Sullivan directed Cedarville University's Center for Bioethics. He still teaches part-time at Cedarville and at Wright State University, and serves as an academic ethicist on two Dayton area ethics committees.

The first associate author of this text is Douglas C. Anderson, Jr., PharmD, DPh, FCCP, professor of pharmacy practice at Cedarville University, Cedarville, Ohio. He received his PharmD from the University of Oklahoma College of Pharmacy in 1992, and completed a residency in pharmacy practice (PGY1) and a fellowship in adult medicine. As a clinical pharmacist, Dr. Anderson's areas of expertise include anticoagulation and thromboembolic disorders, as well as cardiovascular risk reduction. Dr. Anderson has taught bioethics at Cedarville in both the doctor of pharmacy and master of business administration programs. He currently works for the INTEGRIS Health System in Oklahoma City.

The second associate author of this text is Justin W. Cole, PharmD, BCPS, associate professor of pharmacy practice at Cedarville University, Cedarville, Ohio. He received his PharmD from Ohio Northern University in 2006. Dr. Cole worked for 10 years as a pediatric pharmacist at Nationwide Children's Hospital in Columbus, Ohio. There he served as a clinical pharmacist, pharmacy educator, and clinical coordinator, overseeing both clinical and investigational drug services. He now teaches pediatrics and serves as chair of the Department of Pharmacy Practice He also directs Cedarville University's Center for Pharmacy Innovation.

An ethics textbook of this magnitude is a collaborative effort, and we are grateful for a team of colleagues and students who have given us help in bringing it to fruition. The idea for the project began with conversations between one of us (DMS) and Dennis Cooley, professor of philosophy and ethics at North Dakota State University. We sincerely thank Marc Sweeney, former professor of pharmacy practice at Cedarville, and Jeffrey Lewis, former professor of pharmacy practice at Cedarville, and former dean of the Gregory School of Pharmacy at Palm Beach Atlantic University. Both of these scholars had a vision for teaching ethics and law at Cedarville and actively supported our efforts. Kristi Coe, assistant professor of nursing, provided valuable library research support as we began writing.

We owe a huge debt of gratitude to our current and former PharmD students, who enthusiastically embraced this project. In particular, Caleb VanDyke (class of 2019) provided preliminary early research. Rachel Wolthoff (class of 2023) did extensive background research and served as a reader and editor. Jonathan Williams (class of 2021) was deeply involved in this effort and also served as a reader and editor. Jon joins us as coauthor of Chap. 7. Joshua Pearson (class of 2021) was deeply involved as a reader and editor and he joins us as coauthor of Chap. 12. In addition, Josh contributed 5 case studies for Chap. 13.

We are honored to recognize and thank our senior consultant and editor for this book: Jon E. Sprague, RPh, PhD, Director of Science and Research for the Ohio Attorney General and BCI Eminent Scholar of the Center for the Future of Forensic Sciences at Bowling Green State University. Dr. Sprague has graciously written the book's foreword.

In addition, the team at Springer has been wonderful to work with. We especially wish to express appreciation to Dennis Cooley, mentioned earlier, who also serves as series editor for the Springer International Library of Bioethics. We sincerely thank Floor Oosting, Editorial Director for Springer Humanities, and Christopher Wilby, Springer Associate Editor, for their valuable suggestions and overall support.

Finally, Jennifer Sullivan Downey performed extensive proofreading and final editing of the manuscript. We are grateful for the valuable contributions of our readers and editors, but any errors are the fault of the authors.

In conclusion, to all of our past, present, and future students in pharmacy law and pharmacy ethics at Cedarville University, thank you for your enthusiasm and encouragement. We dedicate this book to you.

Cedarville, OH, USA
Dennis M. Sullivan

Douglas C. Anderson

Justin W. Cole

References

Advancing Team-Based Care Through Collaborative Practice Agreements A Resource and Implementation Guide for Adding Pharmacists to the Care Team. Centers for Disease Control and Prevention Web site. www.cdc.gov/dhdsp/pubs/docs/CPA-Team-Based-Care.pdf. Published 2017. Accessed 10 Feb 2021

Reinhart R. Nurses Continue to Rate Highest in Honesty, Ethics. Gallop, Inc. Web site. https://news.gallup.com/poll/274673/nurses-continue-rate-highest-honesty-ethics.aspx. Published 2020. Accessed 10 Feb 2021

Contents

1	**Introduction**.	1
	The Approach to Ethics in This Text.	3
	A Short History of Ethics in Pharmacy Practice.	4
	A Preview of Topics.	6
	References.	8
2	**Basic Ethical Theory**.	9
	Introductory Concepts.	9
	Organizing Ethical Theories.	11
	Ethical Theories Based on Human Reason.	13
	Kantian Ethics.	13
	Natural Law Ethics.	15
	Ethical Theories Based on Tradition.	16
	Divine Command Theory.	16
	Medical Principlism.	19
	Ethical Theories Based on Relationships.	20
	Classical Utilitarianism.	20
	Situation Ethics.	22
	Virtue Ethics.	22
	Summary.	23
	References.	24
3	**Human Value and Human Dignity**.	27
	Human Personhood.	27
	Human Dignity.	32
	References.	34
4	**Clinical Ethics in Historical Context, Part I**.	37
	Medical Principlism and the Hippocratic Tradition.	37
	Informed Consent and Decision-Making Capacity.	42
	Truth Telling in Healthcare.	44
	Pediatric Consent and Assent.	45
	References.	47

5	**Clinical Ethics in Historical Context, Part II**	49
	The Eugenics Era and Its Influence on Modern Healthcare Ethics	49
	The Era of Nazi Medicine	54
	References	60
6	**Clinical Ethics in Historical Context, Part III**	63
	Human Research Ethics: The Tuskegee Syphilis Study	63
	The Willowbrook Hepatitis Studies	67
	Human Subjects Research in the Belmont Era	68
	Key Ethical Provisions of the Belmont Report	69
	The Institutional Review Board	70
	References	73
7	**Pharmacy Professionalism**	75
	The Ethics Code of Pharmacists	75
	Virtue as the Foundation of Pharmacy Professionalism	79
	Health and Wellness Promotion	80
	Everyday Professionalism	81
	Recent Developments	81
	Pharmacy Professionalism and the Law	82
	Conclusion	84
	References	85
8	**Reproductive Ethics**	87
	Contraception	87
	Abortion	91
	Abortion Ethics	94
	Assisted Reproductive Technologies	95
	ART and Surrogacy	100
	Conclusion	102
	References	104
9	**Ethics at the End of Life – Part I**	109
	Introduction	109
	Case Study	111
	Defining Death	113
	The Persistent Vegetative State	116
	Advance Directives	118
	Conclusion	121
	References	122
10	**Ethics at the End of Life – Part II**	125
	The Assisted Suicide Debate	125
	Current Legal Context in the United States	126
	Ethical Arguments Against Assisted Suicide	128
	Ethical Arguments in Favor of Assisted Suicide	132
	Pharmacist Participation in Lethal Injection	135

	Conclusion	139
	References	140
11	**Rights of Conscience**	**141**
	References	149
12	**Vaccines, Resource Allocation, and Unproven Treatments**	**151**
	The Ethics of Vaccines	151
	Ethics During a Pandemic	158
	The Ethics of Unproven Treatments	162
	Pandemic Update	164
	Conclusion	169
	References	171
13	**Case Studies and Policy Scenarios**	**175**
	Clinical Ethics Analysis: Key Steps	175
	What Are the Medical Facts?	176
	What Is the Central Ethical Question?	176
	What Are the Legal and Situational Constraints?	176
	Who Are the Stakeholders, and Who Decides?	176
	What Is the Patient's Quality of Life?	177
	What Religious, Moral, or Cultural Factors Influence the Outcome?	177
	What Ethical Principles Have a Bearing on the Case?	177
	Based on these Principles, What Ethical Duties Conflict, and What Is Their Best Resolution?	177
	What Is the Ethical Conclusion?	178
	Clinical Case Studies	178
	Case Study #1: Dealing with a Difficult Patient	179
	Case Study #2: Dealing with Cultural Beliefs	181
	Case Study #3: The Ethics of Palliative Care	184
	Case Study #4: End-of-Life Decision Making	184
	Case Study #5: Abortion and Malignancy	185
	Case Study #6: Neglect in an 8-Year-Old Boy	186
	Case Study #7: Helping a Patient with Substance Abuse Disorder	187
	Case Study #8: A Patient with a Massive Head Injury	188
	Case Study #9: Should the Pharmacist Dispense this Medication?	190
	Case Study #10: Should This Patient Have Emergency Heart Surgery?	190
	Case Study #11: Medical Marijuana for Chronic Pain	191
	Case Study #12: A Forged Prescription	192
	Case Study #13: Drug Scarcity During a Crisis	192
	Case Study #14: Dispensing a Recalled Medication	193
	Case Study #15: Who Gets the Insulin?	194

Case Study #16: When Should We Stop?	194
Case Study #17: A Handshake Arrangement	195
Case Study #18: Discontinuing an ICD	196
Policy Scenarios	197
Policy Case Study #1: Should the Pharmacist Support This Law?	197
Policy Case Study #2: Midazolam for Lethal Injection	198
Policy Case Study #3: Should We Use Nazi Research Data?	199
Policy Case #4: Rights of Conscience	200
Policy Case #5: Non-Medical Exemptions from a Vaccine Mandate	201
Policy Case #6: Testing a New Drug	202
Policy Case #7: A Grant-Funded Diabetes Study	202
References	203

14 Genetic Ethics and Other Cutting-Edge Issues ... 205
   The Genetic Frontier ... 205
      Gene Editing ... 206
      Three-Person Embryos ... 208
      Genetic Information: Testing and Privacy Issues ... 210
      Precision Medicine ... 212
      Additional Challenges in Genetic Ethics ... 214
   Other Cutting-Edge Issues ... 214
      Machine Learning and Artificial Intelligence (AI) ... 214
      The Rise of Telehealth and Telemedicine ... 215
      The Ethics of Pharmacy Benefit Managers (PBMs) ... 215
      Mandatory Preventative Treatments and Health Behaviors ... 215
      The Continued Opioid Crisis ... 216
   Conclusion ... 216
   References ... 217

Glossary ... 221

Index ... 231

Chapter 1
Introduction

Modern healthcare is becoming more and more complex. Increasingly more sophisticated technology as well as greater specialization among clinicians have led to an environment where the following true-to-life scenarios are common.[1]

1. John W. is in severe pain from a terminal malignancy that has spread widely to his bones. How should he cope with this? His doctors have prescribed high doses of narcotics, but these make him sleep all the time. When he wakes up, his pain returns. His children contact their community pharmacist, Dr. Jessica Eastman, who knows John well. How should Dr. Eastman advise the patient and his family?
2. Dr. Sarah Jameson is a hospital pharmacist asked to give expert testimony before her state legislature. At issue is a proposed law that would limit the use of naloxone (Narcan®) by first responders to resuscitate patients suffering from an opioid overdose. The proposal stipulates that naloxone be used for any one individual a maximum of three times. Supporters argue that the expense of naloxone for small communities is too high and that its repeated use for the same individuals interferes with the care of other patients, such as those with heart attacks, strokes, and trauma, who need urgent transport to emergency rooms. Opponents of the measure worry that such selective use of naloxone is a form of discrimination. How should Dr. Jameson testify? Should she support or oppose this measure?
3. Dr. Jeffrey Rogers works as a pharmacist in a collaborative practice with three physicians. Their clinic deals with the management of chronic pain disorders. Julia S. suffers from the symptoms of fibromyalgia. Though Dr. Rogers and the team have tried different approaches, none of them has eased her pain, and she has become depressed. A friend told Julia how the regular use of marijuana had helped her, so she is anxious to try it. Dr. Rogers realizes that no high-quality evidence supports marijuana for this condition, and the medical or recreational use of the drug is illegal in their state. How should he advise his patient?

---

[1] Though drawn from actual clinical experience, all of these case scenarios are strictly fictitious.

© Springer Nature Switzerland AG 2021
D. M. Sullivan et al., *Ethics in Pharmacy Practice: A Practical Guide*,
https://doi.org/10.1007/978-3-030-72169-5_1

4. Dr. Andreas Marangos is a pharmacist who chairs the Ethics Committee of a mid-sized community hospital. The state attorney general has asked him to approve a contract for compounding a three-drug protocol used in lethal injections for capital punishment. Three inmates are currently on death row in the state prison, with one execution scheduled for the following month. How should Dr. Marangos respond?

All of these stories illustrate the difficulty of balancing patient needs with possible medical interventions, coupled with limitations imposed by law and societal norms. At first glance, these cases may therefore seem incredibly complicated. Some feel that the right thing to do is just a matter of personal opinion, and no real answers exist. But that response is unwarranted, for there are clear ethical principles that inform each of these scenarios. On the other hand, others believe that responding to such cases involves strict adherence to legal and professional standards. That attitude is also misguided, for clinical ethics is more than such a narrow and limited approach. There is, however, a middle ground based on normative ethics and critical thinking, which this book will demonstrate. Our approach emphasizes that there is indeed a best answer in the vast majority of cases. Along with this optimism, we nevertheless place high trust in the clinician's moral judgment. Our goal will be to teach the reader *how* to think, based on ethical principles, not necessarily *what* to think.

To illustrate the strength of this approach, consider the dilemma faced by Dr. Andreas Marangos in Case Scenario #4. Dr. Marangos recognizes that his state's attorney general faces an enormous problem: to carry out a legally-mandated policy of capital punishment without running afoul of the Eighth Amendment proscription against "cruel and unusual punishment." In recent years, lethal injection seems to accomplish this goal by rendering the accused unconscious before injecting drugs to cause death. Note that whether capital punishment is ethical (by American standards of criminal justice) is an issue that extends far beyond the dictates of health care ethics. But should physicians, nurses, or pharmacists participate in some way? The answer to this question is informed by history.

The modern pharmacy profession is only about 150 years old, yet there is a healthcare ethics tradition dating back over two millennia. Pharmacy ethics takes many cues from a binding oath that began in 400 BCE. Hippocrates and his followers were committed to beneficence (having the best interests of patients in mind) and non-maleficence (avoiding harm) (Edelstein 1943; Hulkower 2016). In expanded form, these ideas are part of the modern standard we now refer to as *medical principlism* (Beauchamp and Childress 2019). On this view, the aims and purposes of health care should all be devoted to healing, not hurting. According to the well-known saying, "first of all, do no harm." The Ethics Code of the American Pharmacists Association mirrors the Hippocratic language (Code of Ethics for Pharmacists 2018), and Chap. 4 will cover more details on the Hippocratic tradition.

Against this background, a few concepts concerning capital punishment become more evident. The chemical agents used in a lethal injection cocktail cannot legally or ethically be called *medications* or even *drugs* since they have no clear beneficent purpose. The condemned cannot be referred to as a *patient* where professional obligations might apply. Therefore, a strong line of ethical and legal argument would militate against any participation by Dr. Marangos in the furtherance of lethal injection for capital punishment. As stated earlier, our goal is not to dictate what a pharmacist should think or do, only to recognize the substantial historical precedent that informs this particular scenario (Chap. 10 will further expand on this topic).

The Approach to Ethics in This Text

Because pharmacy ethics is a relatively new discipline, only a few textbooks deal with it. One text (Veatch et al. 2017) uses case studies to inductively and gradually unfold ethical principles after providing a lengthy analysis, and the moral arguments may seem vague and overly theoretical. Pharmacists and PharmD students might be discouraged by the complexity of this treatment. In response to vexing ethical scenarios, the answer provided in most cases is: "It depends on your point of view." This approach offers little practical guidance for the busy clinician.

Another text (Buerki et al. 2013) is a short survey of pharmacy ethics, based primarily on the American Pharmacists Association Code of Ethics. Looking at pharmacy ethics in this way does not adequately focus on ethical history or foundational principles, and does not make clear the distinction between such principles and professional virtues. This treatment may therefore appear somewhat simplistic and overly prescriptive.

In this text, we will navigate between the two extremes of overly theoretical and overly prescriptive, to help the reader learn the task of critical moral thinking. To this end, our framework of normative ethics uses the language of *competing duties*. This idea means that in our analysis of each situation, it will be essential to identify the moral principles that create duties for the clinician. Yet not all apparent duties can be fulfilled within a given clinical scenario.

Consider once again Case Scenario #4. A pharmacist might rightly be concerned that no person should suffer, even while being put to death by execution. Nonetheless, even if this compassionate impulse creates an apparent duty, this is still not correctly a health care professional one. Adhering to the higher burden of professionalism by refusing to provide lethal agents is more central to the traditional role of a pharmacist. Such is the balancing act of normative ethics, and of deciding which duties should prevail. This text presents a clear-cut pathway for resolving ethical dilemmas based on foundational principles and critical thinking.

A Short History of Ethics in Pharmacy Practice

The ethical dimension of medicine has changed dramatically in the years since World War II, and especially in the past thirty years. As cited earlier, the centuries-old Hippocratic tradition was characterized by clinical skill and beneficence, within the moral framework of the doctor-patient covenant. This ethics code, though it began in Greek pagan thought, was readily adapted to a broader context, such that Hippocratic principles became normative for all of medicine (Cameron 2001). Indeed, it is proverbial to say that physicians should not perform actions that would violate their Hippocratic Oath, whether or not they have actually taken such an oath. Though these foundations are being eroded today, the idea of normative ethics in medicine is still compelling and guides much of the modern clinical context.

Pharmacy practice has followed a somewhat different trajectory than medicine. Ethics as a separate discipline is a new topic in pharmacy education since it does not have such a long and hallowed tradition underlying it. Even pharmacy law is itself relatively recent, though now recognized as a critical component of pharmacy training (Fig. 1.1).

INTERIOR OF AN APOTHECARY'S SHOP.
Late XIV. or Early XV. Century.   Flemish.
(From an Old Painting.)

Fig. 1.1 Interior of an apothecary's shop. (Illustration in Public Domain: from Illustrated History of Furniture, From the Earliest to the Present Time from 1893 by Litchfield, Frederick, (1850–1930) – Interior of an Apothecary's shop. Late fourteenth or Early fifteenth century. Flemish. (From an Old Painting))

Early community pharmacists were first called *apothecaries,* who prepared and dispensed remedies and offered front-line medical advice. At first, there was almost no standardization and no regulation for medication purity, quality, or counterfeits. Apothecaries became *pharmacists* about 150 years ago, and with this change in designation came much more legal scrutiny (Anderson 2001).

The first official U.S. Pharmacopeia (USP) began in 1820 (USP Timeline 2018). During this time, unregulated patent medicines offered cures for a wide variety of ailments, and merchants often did not know much about pharmaceutical chemistry or medicine. This deficit came to the attention of the public during the Mexican-American War of 1846 to 1848, where many soldiers died from adulterated drugs. After a public outcry, the Drug Importation Act of 1848 addressed counterfeit and adulterated drug imports, using chemists to analyze and inspect imports, with the USP as a reference for characterizing and logging medications. Nonetheless, enforcement of the Drug Importation Act was ineffective and short-lived, and food and drug adulteration problems continued. These problems ultimately led to the Pure Food and Drug Act of 1906. However, genuinely effective government regulation would not occur until the Food, Drug, and Cosmetic Act of 1938, which formed the basis of most of the modern pharmaceutical laws and regulations in the United States (Abood and Burns 2017).

It is therefore worth noting that "jurisprudence" first became a national standard in pharmacy education with the publication of *The Pharmaceutical Syllabus* in 1910, although it appears that implementation in pharmacy curricula was on a state-by-state basis.[2] Since that time, pharmacists have been considered experts in law, much more than physicians or nurse practitioners. Pharmacists need such expertise, for they are the last layer of protection from mistakes in drug choice, dosing, adverse drug interactions, adulteration, misbranding, and a host of other potential patient harms. Today, pharmacists must also be experts in ethics. While ethics and law complement each other, they are distinct fields. Ethics is more important than ever in pharmacy practice, for two principal reasons.

First, the modern practice of pharmacy has become highly complex. For example, technological advances have led to the emerging arena of *precision medicine.* This approach treats diseases in a customized fashion based on the genetic and environmental factors of each patient. The past few decades have seen an explosion of new and highly technical approaches that pharmacists must master.

Second, the modern pharmacy practice environment is changing rapidly. In the past, the pharmacist had a more subordinate position, with physicians, dentists, and nurse practitioners (not to mention veterinarians, podiatrists, and others) doing the major diagnostic decision-making and prescribing. The pharmacist's duty was to provide support and to carry out the professional dictates of first-level providers. Now with new trends towards provider status for pharmacists, health care has focused more on patient-centered care. This interdisciplinary approach is all about

---

[2] Our thanks to William E. Fassett, PhD, RPh, FAPhA, Editorial Director, American Society for Pharmacy Law for this insight.

the healthcare "team," where pharmacy as a component discipline holds increasing importance. Ethical deliberations are a vital part of this expanded role. This need is now well-recognized within the profession, and accreditation standards for pharmacy training programs include the ethics of health care delivery as an essential component.[3]

In the pages that follow, each chapter of this book will discuss the history and philosophical background of many ethical issues in detail. We will give you the tools to deeply analyze each of the case scenarios presented earlier, along with many others, and we will show you a clear-cut pathway for resolving each one.

A Preview of Topics

Section I of our text deals with *Foundations*. You are reading the introductory chapter, which will be followed by Chap. 2, covering fundamental ethical theories, and Chap. 3, a discussion of different approaches to human value. At first glance, the reader might be tempted to skim over these chapters or omit them altogether. That would be a mistake, for their content is not overly complicated or philosophical, and both contain practical examples.

Though some approaches look directly to medical principlism or the American Pharmacists Association (APhA) Code of Ethics, it will be essential to consider those sources against a backdrop of moral filters or lenses. Chapter 2 provides such a context, along with the foundational language of ethical theory that will help the clinician do the critical-thinking balancing act that moral judgments demand.

Chapter 3 shows the importance of clear and accurate thinking about human persons as members of the moral community. We will present two possible ways of understanding human value, as well as two competing definitions of human dignity. This approach will help the reader to better understand the historical controversies that these ideas inform.

Chapter 4 begins a historical survey of clinical ethics. We will start with the Hippocratic tradition, an ancient code of medical ethics that first articulated the precepts of beneficence, non-maleficence, and distributive justice. Much later, the eighteenth century principle of autonomy was added, leading to our modern idea of medical principlism. The chapter will then discuss concepts of competency and decision-making capacity (not the same thing), and how they contribute to our current understandings of informed consent. These ideas are further refined by their application to vulnerable special populations, such as children, prisoners, and the physically or cognitively disabled. We will conclude with a discussion of truth-telling and confidentiality.

---

[3] PharmD Program Accreditation. Accreditation Council for Pharmacy Education Web site. www.acpe-accredit.org/pharmd-program-accreditation/. Published 2020. Accessed 2/15/2021.

Chapter 5 should remind us of the famous aphorism attributed to philosopher George Santayana: "Those who cannot learn from history are doomed to repeat it" (Santayana 1911). We will begin with a discussion of the tragic pseudoscience of eugenics, which had a significant influence on nineteenth and early twentieth century social theory. However, it also profoundly influenced the Third Reich, leading to research abuses against Jews and others by Nazi doctors during World War II. We will also discuss the fascinating question of whether the research data obtained should ever be used in the future.

Chapter 6 will consider further departures from Hippocratic principles that followed after the war. The flawed philosophy of eugenics also contributed to the racism of the Tuskegee syphilis studies at the beginning of the Great Depression. There were a significant number of other human research violations in the twentieth century, leading eventually to the need for the Belmont Code of 1979, which served as a foundation for modern institutional review boards.

Chapter 7 is all about pharmacy professionalism and begins by showing how the modern APhA Code of Ethics derives many of its principles from the Hippocratic tradition. We will also examine how virtue ethics may be a strong foundation for the ethics of pharmacy practice, and give several examples of what this might look like in practical application.

Having considered the foundations of pharmacy ethics, Section II of our text deals with *Issues and Cases*. We will begin this section in Chapter 8, with a deep dive into reproductive ethics, including contraception and abortifacient drugs. We will also consider the difficult subject of abortion, both in constitutional law and on both sides of the ethical debates. Finally, we will examine the newer terrain of assisted reproductive technologies.

Chapter 9 of our text will be the first of two on ethics at the end of life. We will begin with an examination of the parallel ideas of hospice and palliative care. Next, we will use a sample case study to define several issues surrounding end-of-life ethical concepts, including when to limit or withdraw certain treatments. We will also consider the varying definitions of cardiopulmonary death, brain death, and the persistent vegetative state, along with their ethical considerations. We will conclude the chapter with a discussion of advance directives.

Chapter 10 is devoted to one of the most substantial debates in modern healthcare ethics, that of assisted suicide. We will carefully examine both sides. We will then conclude with an analysis of pharmacists and their possible role in lethal injection for capital punishment, a topic we introduced earlier in this introduction.

Chapter 11 will take up the complicated matter of conscience rights in healthcare. What is the status of sincerely-held personal opinions in pharmacy practice? Specifically, should pharmacists have the option to decline to dispense specific medications that run counter to their own religious and ethical beliefs? Does it matter if such opinions are backed up by medical evidence? What role does the idea of moral complicity play?

Chapter 12 will deal with additional ethical issues in pharmacy practice, including controversies over vaccines, how to respond ethically during a pandemic, and the ethics of unproven treatments.

Chapter 13 is all about case studies and policy scenarios, designed to put into practice what you have learned in this text. We will begin with a specific approach to case study analysis that will help the clinician to balance competing ethical duties. We will analyze and give guidance in two of these cases but will leave the majority for you to consider on your own. The cases derive from the various specific disciplines of pregnancy care, pediatrics, cardiopulmonary conditions, renal diseases, neurological and psychiatric disorders, and cultural and religious issues. We will conclude with several tough questions from a broader policy perspective.

Finally, Chapter 14 considers genetic ethics and other cutting edge issues. We will examine the controversies surrounding human gene editing, three person embryos, and genetic testing. Finally, we will list several challenges that have not yet received an adequate ethical analysis.

Applied Pharmacy Ethics A Practical Guide is intended as a useful and informative resource to help both students and busy practitioners. Each chapter will contain a glossary, questions for review, and recommendations for further reading, plus notes and references. The end of the text will feature a complete glossary and a comprehensive index. Along the way, be sure to master definitions, which will go a long way towards providing clarity on challenging or subtle concepts.

Let's get started.

References

Abood RR, Burns KA. Pharmacy practice and the law. 8th ed. Burlington: Jones & Bartlett Learning; 2017.

Anderson S. The historical context of pharmacy. In: Taylor K, Harding G, editors. Pharmacy practice. Boca Raton: Taylor & Francis; 2001.

Beauchamp TL, Childress JF. Principles of biomedical ethics. 8th ed. New York: Oxford University Press; 2019.

Buerki RA, Vottero LD, American Pharmacists Association. Pharmacy ethics: A foundation for professional practice. Washington, DC: American Pharmacists Association; 2013.

Cameron NM. The new medicine: life and death after Hippocrates (revised ed.). London: Bioethics Press; 2001.

Code of Ethics for Pharmacists. American Pharmacists Association. www.pharmacist.com/code-ethics. 2018.

Edelstein L. The Hippocratic oath, text, translation and interpretation. Baltimore: The Johns Hopkins Press; 1943.

Hulkower R. The history of the Hippocratic Oath: outdated, inauthentic, and yet still relevant. Einstein J. Biol. Med. 2016;25(1):41–4.

Santayana G. Introduction and reason in common sense, vol. 1. New York: Scribner; 1911.

USP Timeline. US Pharmacopeia. www.usp.org/about/usp-timeline. 2018.

Veatch RM, Haddad AM, Last EJ. Case studies in pharmacy ethics. 3rd ed. Oxford/New York: Oxford University Press; 2017.

Chapter 2
Basic Ethical Theory

Introductory Concepts

Any study of professional ethics must begin with the basics, so this chapter will open with the question, "What is ethics?" The word *ethics* is a generic term that could mean many things but generally refers to the study of principles of right and wrong behavior (Fieser 2018). A synonym for ethics is *moral philosophy*. As a discipline, moral philosophy is broken down into three branches: meta-ethics, normative ethics, and applied or professional ethics. *Meta-ethics* is the study of how moral philosophy is defined, examining such moral concepts as right, wrong, justice, etc. *Normative* ethics consists of commonly-held principles or theories that can be used by everyone. *Applied or professional* ethics takes ethical principles and puts them to work. In our case, this will be the realm of pharmacy ethics (Fieser 2018).

In common usage, people often throw around the words *ethics* and *morality*. These terms are closely related, with the latter referring to a specific set of norms or principles by which people live. So when we think of morality, we commonly think of standards derived from religious or philosophical precepts, while ethics usually refers to more specific "rules of the game," applied to behaviors of particular groups. In actual practice, however, many ethicists use the two terms interchangeably. So, for example, if we say that something is unethical, we also mean that it is immoral. This is a typical practice and will be true of our discussions as well (Feinberg and Feinberg 2010).

Since it involves human behavior and its rules, ethics is entirely different from other disciplines. As an example, clinical pharmacy practice is an empirical science. Standards by which certain antibiotics are used to treat community-acquired pneumonia are arrived at through clinical research and observation. On the other hand, moral philosophy is a different sort of discipline, consisting of a series of *behavioral recommendations*. Instead of using the empirical word *is*, ethics employs the term *ought*. In other words, empirical statements are *declarative* (in the grammatical sense), and ethical statements are *imperative*.

So this raises the interesting question, "Is moral philosophy objective? Are there normative standards that define the rules for all human agents, or is this merely a matter of subjective preference or social consensus?" In this text, our approach is based on normative ethics, the branch of moral philosophy that considers and defends commonly-held frameworks for ethics. From this perspective, we will make recommendations derived from universally agreed-upon principles. These standards consist of widely-shared norms about right and wrong human conduct and form a stable social consensus. These will be especially true of healthcare ethics, where the rules are not particularly controversial or subject to debate, even though they may be incomplete. In short, normative ethics is the branch of moral philosophy that applies to everyone. "Morally serious" persons accept these ideas as authoritative. Some examples of this broad consensus include: do not lie, do not steal, do not kill, and do not commit incest.

Professional ethics is a subset of normative ethical theory that is binding on a particular professional group. Each profession has its own set of guidelines, e.g., law, engineering, teaching, and health care. *Pharmacy ethics* is a further subset of health care professional ethics, dealing with specific issues that might not be relevant to other disciplines. These matters might include such things as informed consent, confidentiality, and autonomy.

> **Three Branches of Moral Philosophy**
>
> Meta-Ethics: defines ethical concepts
> Normative Ethics: commonly-held principles (ethics for everyone)
> Professional Ethics: applies ethical principles (e.g., pharmacy ethics)

Before addressing individual ethical theories, it is essential to understand how the "oughts" of moral philosophy function in human experience. Because ethical duties are moral judgments, failing to live up to these may bring a real or perceived sense of moral criticism or guilt. The imperative nature of ethical standards of conduct implies *freedom of action*. This idea means that free moral agents are able to choose (Feinberg and Feinberg 2010). A short example will clarify.

> A 6-year-old child is drowning in a nearby pond. You are hiking nearby and hear her cries for help. What should you do? Most would agree that you have a duty to jump into the water, swim out to the child, and save her life. You could choose not to do so, but this would be morally wrong.
>
> But what if you don't know how to swim? While you plausibly might be able to render assistance in another way, the child will likely die by the time someone else comes along. Have you therefore committed a moral wrong? Does your failure to act invoke moral censure? While a strong sense of regret or perceived guilt is inevitable, most would intuitively agree that your failure to act in this instance is a tragedy, but not a moral failure.

Another implication of the "freedom of action" principle in ethics is the idea that there are three possible levels of moral duty involved with every human action:

1. Morally neutral: actions with no ethical import. Examples might include the choice of color for one's clothes, preference for coffee over tea, or the time one goes to bed at night.
2. Morally obligatory: actions that must be done or must be avoided. Failure to fulfill these duties brings moral blame. Examples might include deliberately causing avoidable harm or lying for selfish personal gain.
3. Morally heroic: actions that exceed minimal obligations, that go "above and beyond" mere duty. One example from the medical realm might be donating bone marrow or even a kidney to a total stranger. Such an act, while not morally required, is universally recognized as noble and sacrificial (Feinberg and Feinberg 2010).

Organizing Ethical Theories

To organize our discussion of theories, let us begin with a classic ethical thought experiment, the Trolley Problem. British philosopher Philippa Foot first proposed this illustration in 1967 (Fig. 2.1) (Foot 1967, 2002).

Our version of this dilemma is as follows:

> You are standing at a distance from a railroad yard, but close enough to see the tracks. You notice a trolley bearing down on an unsuspecting group of five people. You are too far away for them to hear your warning, but you are right next to a lever that controls a switch. Changing the points on the switch would divert the trolley onto a side track, avoiding the five bystanders, but killing another innocent person on the side track. Would you throw the lever and divert the trolley?

Many people would instinctively vote to throw the lever, reasoning that it is better for only one person to die than five. But let us change the scenario slightly, as follows:

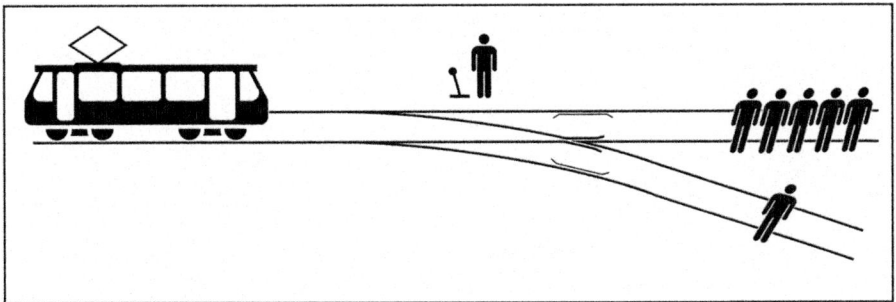

Fig. 2.1 The Trolley problem. (Illustration by McGeddon – Own work, CC BY-SA 4.0, https://commons.wikimedia.org/w/index.php?curid=52237245)

You are now standing on a bridge over the railroad yard, and you notice a trolley bearing down on a group of five people. You are still too far away for them to hear your warning, but a rather heavy-set man is standing next to you. You have the option of pushing him onto the track, reliably stopping the trolley and saving the five unsuspecting bystanders, but at the cost of killing this one man. Would you do this?

Many who voted to throw the lever in the first instance would reject pushing the man onto the track in the second, even though the outcome in both cases is the same. Why is there a difference? The answer illustrates two major kinds of ethical theories. The first instance is a demonstration of *consequentialist* reasoning. Consequentialist ethical rules rely on the results as the sole determinant of right and wrong. Here, the calculus of allowing one person to die to save five others makes sense. However, the second instance illustrates the effect of *deontological* reasoning. Deontological ethical theories rely on core principles, irrespective of the outcome. In this case, the idea of harming a man who was not previously threatened, even to save others, violates strong societal strictures against intentionally causing death. So the case illustrates one basic approach of classical ethics, dividing theories into two categories: consequentialist and deontological (Fieser 2018; Feinberg and Feinberg 2010; Shafer-Landau 2018).

To further illustrate the two categories, let us return to the Trolley Problem, this time re-framed in the context of health care:[1]

The chief of the transplant team at a major medical center has a genuine dilemma. Five of his current patients are now in the hospital, desperately ill, each in need of a different organ. Without the transplant, each patient is sure to die. That same day, a healthy young man comes in for a routine checkup. The doctor examines the young man, discovering that he is a tissue match for all of his desperately ill patients. If he could get away with it, should the doctor kill the young man for his organs to save five other lives?

Based on clear-cut moral principles, the answer, of course, is no. The deontological case against such an act is strong. But now let us modify the scenario (as we did with the Trolley Problem), to see if our response might change:

As before, the chief of the transplant service has five desperately ill patients in need of a transplant. A previously healthy young man is involved in a terrible motor vehicle accident and suffers a massive head injury. A careful and thorough clinical evaluation determines that the man is brain dead and that he is a tissue match for all of the patients who need a transplant. Assuming that all proper legal conditions are met, is it ethical to remove organs from the young man to save the five patients?

This altered scenario changes everything. While there might be many hidden nuances to a thought experiment such as this one, it seems plausible that the transplant surgeries could ethically proceed, based on consequentialist reasoning.

The deontological and consequentialist categories often provide the framework for discussing ethical theories, dividing them into one or the other. The problem with this approach is that not every theory fits neatly into one of the two categories; some incorporate elements of both. For this reason, we will propose a different framework by which to categorize ethical rules:

---

[1] Our medically-oriented versions of the trolly problem are based on an idea from Thomson JJ. The trolley problem. *The Yale Law Journal*. 1985;94(6):1395–1415.

1. Reason
2. Tradition
3. Relationships

Reason-based systems rely heavily on human rationality, tradition-based systems depend on received wisdom from the past, and relationship-based systems center around social interactions with others. Along the way, the terms deontological and consequentialist will also help to describe the characteristics of each theory.[2]

Why study multiple ethical theories? Ethical reasoning depends on the balancing of competing duties, and these duties arise from numerous points of view. Healthcare professionals, whether physicians, nurses, therapists, dietitians, or pharmacists, come from a wide variety of backgrounds, beliefs, and worldviews. Normative ethics holds to the optimistic idea that such professionals can arrive at a reasonable resolution to an ethically complex case despite their different backgrounds. We should all learn the various ethical theories, including their strengths and weaknesses, and become good at utilizing every one of them. In other words, we should become "fluently multilingual" in ethics. Only then can we properly engage difficult cases from a normative approach. With all this in mind, let us consider the theories.

Ethical Theories Based on Human Reason

Kantian Ethics

Case Study
Elaina Fletcher is an 82-year-old woman who recently lost her husband of 56 years to cancer. She has been depressed in recent weeks, but with medication and counseling, she is slowly starting to enjoy life once again and is planning to go on a tropical cruise next month with a close friend.

Dr. Robert Anton is Elaina's internist, who has taken care of her for many years. Because she had been complaining of mild low back pain, he performed a series of tests, including a CT scan of the abdomen. The results are not good. A large mass is present in the posterior abdominal region, pressing outward onto lumbar spinal nerve roots. The diagnosis is almost certainly an incurable malignancy of the pancreas.

Dr. Anton is tempted to withhold this information from Elaina. He reasons that she is already depressed and that this news would be devastating to her. She is also planning a long-awaited vacation, which she deserves to enjoy. His first impulse is to lie to his patient and tell her that everything is fine, at least until she returns from her cruise.

---

[2] Our three-part grid for organizing theories came originally from Feinberg JS, Feinberg PD. *Ethics for a brave new world*. 2nd ed. Wheaton, Ill.: Crossway; 2010. We have modified the authors' original format to better cover ethical theory, especially as applied to health care.

Ethics Question

Should Dr. Anton lie to Elaina?

Immanuel Kant was an eighteenth century German philosopher who developed his philosophical ideas based on pure reason, apart from religion. His ethical theory presupposes a morally serious person, bound to do what 'duty' demands. As we have seen, ethics consists of imperative statements in the form of recommendations for proper behavior. Kant teaches that there are two types of these statements: *hypothetical imperatives* and *categorical imperatives* (Johnson and Cureton 2016; O'Neill 2009).

A hypothetical imperative is a simple if/then statement, a conditional command. Here is an example: "If you want to pass the ethics exam, then you should study the material." Note that this is just a matter of good judgment to help someone get ahead, which Kant would call mere prudence. There is nothing particularly ethical about it; it is just a formula for success.

On the other hand, acting ethically depends on Kant's categorical imperative. According to the first and best-known version of this, we should perform an action only if we could potentially make it a binding rule for everyone (Kant calls this a 'maxim'). Let us try out the rule on our example. Based on sympathy for his patient, Dr. Anton considers the possibility of lying to her. He makes it into a maxim: 'doctors should always lie to their patients.' It seems intuitively obvious that this is unsustainable, since following the maxim would destroy the trust relationship between doctors and their patients and ruin the practice of medicine. Therefore, since Dr. Anton cannot defend lying in the general case, he cannot defend it in this specific instance. He must tell Mrs. Fletcher the truth, even if this is difficult.

Notice that this approach to ethics is deontological, based on principles derived from reason that admit of no exceptions. And yet it also has a somewhat consequentialist flavor to it, since one must look to theoretical outcomes in examining the categorical imperative.

Kant also posits a lesser-known but useful additional rule. His second categorical imperative states that persons should be ends in and of themselves, never the means to another person's ends.[3] He thus argues against "using" people, and this idea has been influential in the later development of ideas about human rights (Johnson and Cureton 2016; O'Neill 2009).

Kantian ethics has several strengths. It accords with most people's ideas about moral right and wrong. It implies an objective standard of moral truth, making us less prone to rely on our feelings and emotions. However, its basis is an almost cold commitment to moral duty. Without appealing to deity or other authority, many critics have questioned if there is any motivation for "morally serious" persons to act in this way.

Kantian ethics also cannot handle moral conflicts (Wilkens 2011). As we have seen, analyzing ethical cases consists of balancing competing ethical duties. When

---

[3] Those familiar with Judeo-Christian concepts will note that this sounds a lot like the Golden Rule, which states, "Do unto others as you would have them do unto you." The Christian Bible gives this principle in Matthew 7:12.

there are two or more duties to be obeyed in a specific situation, the first type of categorical imperative may have limitations in providing the necessary insight. Kantian ethics admits of no exceptions.

So should Dr. Anton lie to Elaina Fletcher? According to Kant, the answer is no.

Natural Law Ethics

Case Study
Jonathan Andrews is a 38-year-old male insurance executive, who was driving to his office early one morning. He was in a hurry, juggling his coffee and his cell phone, and neglected to fasten his seat belt. Therefore, Jonathan didn't see the massive tractor-trailer pulling in front of him until too late. He struck the truck a glancing blow, which deployed his airbag. The car then veered off the road and crashed into a tree, throwing Jonathan from the vehicle. Observers called 911, and an ambulance took him urgently to a nearby emergency room.

In the ER, Jonathan was found to have a massive head injury, with a dangerously elevated intracranial pressure. He was deeply comatose. After evaluation in the ER by the neurosurgeon on call, he underwent CT scans and pressure monitoring. He eventually was stabilized in the neurosurgical intensive care unit.

It is now two months later, and Jonathan is in a persistent vegetative state (PVS). He can breathe on his own but has no conscious interaction with his environment. He has a feeding tube in place to maintain nutrition and hydration.

Ethics Question
Can the feeding tube be removed, or is continuing the feeding tube morally obligatory?

Thomas Aquinas was a thirteenth century medieval theologian and philosopher, recognized as a Church Father in the Roman Catholic Church (Brown 2018). John Locke, a seventeenth century British political philosopher, was a strong influence on Thomas Jefferson and the writing of the Declaration of Independence (Connolly 2020). Both Aquinas and Locke articulated ideas we now refer to as *natural law*. Natural law is both a philosophical framework and a moral theory (Murphy 2011). In our discussion, we will focus on its ethical implications.

To better understand the core principles of natural law, consider an excerpt from the Declaration of Independence: "We hold these truths to be self-evident, that all men are created equal, that they are endowed by their Creator with certain unalienable Rights, that among these are Life, Liberty and the pursuit of Happiness."[4] Jefferson here appeals to a moral standard accessible purely from reason, i.e., it is self-evident. He is talking about rights that arise from the way a divine God designed human beings. Jefferson believed that because of our created human nature, we have certain "unalienable" (absolute, non-negotiable) rights.

---

[4] The Declaration of Independence. *US History.* 2018. www.ushistory.org/declaration/document/.

The foundation of natural law as a moral theory is the claim that morality can be derived from the nature of the universe and the nature of human beings. Its principles arise from a rational understanding of truths common to all moral agents. This whole idea of embedded truths is inherently theistic, since, as the Declaration implies, these were built in by a Creator. On this view, there is nothing arbitrary or subjective about such knowledge (Budziszewski 2011).

Using natural law as a basis for ethical decision-making may be difficult, however, because not everyone will agree on what is "natural" or "built-in" as a part of human nature (Feinberg and Feinberg 2010; Wilkens 2011). Nonetheless, such claims are common. As one example, many people are suspicious of hormonal birth control as going against the way God or nature intended for procreation to take place. This theme is very prominent in the Magisterium (teaching authority) of the Roman Catholic Church. Another example is the intentional killing of a human person, even if that person requests it in a medical context (as in euthanasia). Many would claim that such an act violates a built-in instinct for survival fundamental to human nature (note that euthanasia also runs counter to several other ethical theories).

As a moral theory, natural law may provide a helpful perspective for those who just don't "feel right" about specific actions. It correlates with the instincts of a significant element of the population, especially those from a religious background. Yet it may also seem largely intuitive and subjective. Opponents of natural law argue that we cannot go from empirical facts to values, i.e., we cannot go from an "is" to an "ought." They would claim that this represents the "naturalistic fallacy," (Duigna 2012) and that this is a logical error. Nonetheless, this is an "error" that many ethicists make, and natural law continues to exert a great deal of influence.

What about Jonathan Andrews, the patient in PVS? Though severely handicapped, he does not meet the standard criteria for brain death. Those who argue against discontinuing nutrition and hydration may make their argument by using natural law. Providing food and water is part of the minimal ethical duty we owe one another, they will say, since it is normal and natural and required for survival. This claim is, of course, not without controversy. Similar very public cases, such as those of Karen Ann Quinlan, Nancy Cruzan, and Terri Schiavo, have all generated intense legal and ethical scrutiny, and are still passionately debated today (Chap. 9 discusses these cases in detail) (Campbell 2017).

Ethical Theories Based on Tradition

Divine Command Theory

Case Study
Sandra Stockman is a 42-year-old mother of two children, ages 2 and 5. She has recently discovered that she is again pregnant, now in her twelfth week of gestation.

At her age, Sandra had been concerned about the possibility of fetal defects, so she asked her obstetrician to perform an amniocentesis. This prenatal test samples the amniotic fluid to look for chromosomal abnormalities.

The results were upsetting. Dr. Susan Walters revealed to Sandra that her unborn fetus had Trisomy 21, the genetic marker associated with Down syndrome. Without much discussion, Dr. Walters told Sandra that her best option was a medical abortion. She gave her a single pill of mifepristone (200 mg.) and a prescription for eight tablets of misoprostol (200 mcg.), with instructions to take the mifepristone that evening, then four of the misoprostol tablets 48 h later.

Sandra is distraught over this advice. She and her husband, Jeff, stop in at their small town's only pharmacy, where Dr. Ben Wilson is on duty. Sandra and Jeff have known Dr. Wilson for years, and they now ask for his advice. In a private consulting room, Sandra explains to the pharmacist that she is Roman Catholic. Though she and Jeff don't often go to church, they plan to raise their small children in the faith. The idea of abortion violates clear teachings from their priest, but they also worry about bringing a disabled child into the world.

Ethics Question

How should Dr. Wilson counsel this couple?

The divine command approach to ethics is based on the meta-ethical notion that moral obligation comes directly from a deity or deities. Also called theological voluntarism (*voluntas* is Latin for *will*), in divine command theory (DCT), God's *will* determines right and wrong (Feinberg and Feinberg 2010). There are many variations of DCT, each derived from a different authoritative source, e.g., Christian biblical ethics, Koranic ethics, Jewish ethics, etc. Our broad analysis will address DCT from a monotheistic, Judeo-Christian perspective.

A favorite bumper sticker for this approach declares, "God says it, I believe it, and that settles it" (Wilkens 2011). This slogan implies a firm position that allows for no dissent and can be overly simplistic. The aphorism "God said it" could just be an excuse for poor reflection, or an attempt to justify one's prejudice. Nonetheless, there is enormous and complex literature on such theistic ethical theories, with some going back thousands of years.

Many healthcare professionals are understandably uncomfortable with DCT, but for most people, religion and ethics are inextricably intertwined. In a pluralistic context, religiously-informed views are relevant to ethical discussions. Those who hold to such viewpoints should not automatically assume that others will agree with them, but these ideas should have a seat at the table. With that in mind, divine command ethical views have some apparent strengths and weaknesses.

One advantage of DCT is that it helps avoid the problem of relying too heavily on flawed human reason. There is a certain humility implied by seeking guidance from an authority outside ourselves. Borrowing from the traditions of a church, synagogue, temple, mosque, or other religious authority allows for shared decision-making with like-minded counselors, and this can bring depth, meaning, and comfort to a suffering patient and the patient's family (Feinberg and Feinberg 2010).

There are, however, some potential problems, including a fascinating classical objection to DCT dating back to Plato. In one of the Socratic dialogues, the great teacher Socrates poses a fundamental question to his colleague Euthyphro: "The point which I should first wish to understand is whether the pious or holy is beloved by the gods because it is holy, or holy because it is beloved of the gods."[5]

For our purposes, we can eliminate the polytheism and convert this into a question of ethics: "Is something right because God commands it, or does God command us to do what is right because it is right?" As one writer put it, is God a legislator or a Supreme Court justice? If God's commands are merely the result of legislation, then this seems arbitrary since God could decree the opposite, and we would have to obey. On the other hand, if there were a pre-existing standard to which God conforms, then he would not be an omniscient and omnipotent being, as the idea of deity typically requires (Wilkens 2011).

Many moral philosophers have considered the so-called Euthyphro's Dilemma a severe problem for DCT. However, people of faith who believe in an omniscient and omnipotent deity would merely posit that God is *both* legislator and Supreme Court justice, that his eternality into the past demonstrates that he both determines the moral standard and exemplifies it (Austin 2018). This response resolves the apparent dilemma to their satisfaction.

Why is this discussion important for modern-day applications of DCT in ethics? A medieval theologian named William of Ockham held that human acts are not intrinsically right or wrong, but only moral as commanded by God, and not accessible to reason (Austin 2018). This idea implies that God could change His mind about the Ten Commandments and that we would then be obligated to obey their opposites. Some variants of his extreme DCT are still around today, leading to a blind, non-rational form of theological legalism. It is not uncommon to see some religious approaches that take portions of sacred texts out of their cultural and historical contexts, leading to conclusions never intended by the original writers.[6] Modified DCT tries to avoid this problem by claiming that no actions are morally arbitrary, that morality is embedded in the nature of God and of the universe (note the affinity here with natural law). Ethical rules written in sacred texts are, therefore, grounded in God's very nature. They can thus serve as helpful guides if not overly or legalistically interpreted (Wilkens 2011). Since there are many religions and sources of sacred wisdom, the fine-tuning of DCT approaches for ethical guidance remains controversial.

Based on a modified understanding of DCT, how should the pharmacist Ben Wilson counsel Sandra Stockman and her husband? He recognizes that abortion is a legal procedure and that many health professionals are willing to consider a first-trimester abortion if the fetus has a chromosomal abnormality such as Trisomy 21.

---

[5] Lamb W. Plato's Euthyphro, Apology, Crito-Plato's Euthyphro, Apology of Socrates, and Crito. Edited with notes by Burnet John. Oxford: Clarendon Press, 1924. 8s. 6d. net. *The Classical Review.* 1924;38(7–8):203–204.

[6] Slick M. What is Legalism? Christian Apologetics & Research Ministry Web site. https://carm.org/what-is-legalism. Published 2018. Accessed 2/15/2021.

However, Dr. Wilson also understands that the Stockmans hold to a religious worldview that considers termination of pregnancy morally illicit. They are conflicted about all this. A well-informed professional will ensure that the Stockmans understand the risks and benefits of all the options available to them, including the challenges of raising a child with mild cognitive disabilities and other possible physical problems (Jonsen et al. 2015). Note that even if Dr. Wilson does not share the Stockmans' particular moral views, he can still carefully counsel them, respecting their religious sensibilities.[7]

Medical Principlism

Since this text deals with the ethics of pharmacy practice, it should be clear that this is a relatively new endeavor. As we saw in Chap. 1, the profession of pharmacy is only about 150 years old, yet there is a healthcare ethics tradition that dates back many centuries, in fact, over two millennia. Pharmacy ethics rightly takes many cues from the rich tradition that comes from Hippocrates, dating back to 400 years BCE (Edelstein 1943).

Hippocrates and his followers developed a more rigorous and scientific understanding of medicine that was revolutionary for its day. In accordance with a binding oath, Hippocratic practitioners were committed to beneficence (having the best interests of patients in mind), non-maleficence (avoiding harm), and distributive justice (treating all patients equally, regardless of gender, social class, or other medically non-relevant factors). More recently (in the eighteenth century), the principle of autonomy was added, making up four significant ideas we now refer to as *medical principlism* (Jonsen et al. 2015; Beauchamp and Childress 2019).

Because of its overriding importance for healthcare, we will more fully discuss the Hippocratic tradition and medical principlism in Chap. 4. We insert these concepts here merely to point out that they fit into our three-part schema under the broad category of tradition-based theories. As an example of received wisdom from the past, medical principlism is deontological and duty-oriented in its ethical character (Cameron 2001). As far as Hippocrates is concerned, note that there is another example of this sort of medical moral tradition, based on the Oath and Prayer of Maimonides, a twelfth century Jewish scholar and physician. Overall, these ideas are not as well-known, but they reflect similar duties of beneficence, non-maleficence, and justice[8] (Veatch 1995).

---

[7] The ethical issues surrounding induced abortion are very complex, yet important for pharmacists. These matters will be addressed in greater detail in Chap. 8.
[8] Prayer of Maimonides. *California State Journal of Medicine.* 1918;16(1):51–51.

Ethical Theories Based on Relationships

Classical Utilitarianism

Case Study
Dr. Alison Davies is a hospital-based pharmacist in Unity, Vermont,[9] who works for a healthcare system that includes two clinics and three hospitals, one of them a pediatric facility. Unity is a medium-sized community in the middle of a challenging epidemic of influenza, one that was not well anticipated by the Centers for Disease Control and Prevention (CDC). This particular viral strain has proven to be especially problematic for children, and two children in Unity have already died this year from the disease. Several adults have also been hospitalized, but they were successfully treated and have recovered.

Unfortunately, because this particular virus was unanticipated by the CDC, the vaccine specific for the strain is in very short supply. Dr. Davies, out of consideration for the children at the highest risk, has developed a proposal for consolidating the remaining supply of the vaccine within the area. She proposed purchasing it from local pharmacies and preferentially using it to vaccinate children. This plan would prevent clinics and pharmacies from having any access for adults, some of whom will be at high risk, but would improve the situation for children. Dr. Davies wants to use the remaining limited supply of the vaccine where, in her professional opinion, it will do the most good. She is, however, worried about the ethical and legal implications of the idea, and has submitted her proposal to the Hospital Ethics Committee for its input.

Ethics Question
Should the Ethics Committee approve this plan?

Jeremy Bentham was an eighteenth century British philosopher and social reformer who developed a consequentialist theory of ethics known as utilitarianism. Classical utilitarianism derives from Bentham's Principle of Utility, which sought to maximize pleasure among moral agents. One shorthand way of expressing this idea is as follows: "The morally right action is that which maximizes pleasure for the greatest number of people." This formula thus applies only to results. Bentham himself was opposed to religious ideas and felt that there were no overarching principles to guide human behavior, only outcomes. He even tried to "quantify" these deliberations, developing a hedonistic (pleasure-based) calculus, wherein one could calculate the right action through a detailed equation (Sweet 2018).

John Stuart Mill was a British philosopher and economist who wrote primarily in the nineteenth century (Macleod 2016). As a former student of Bentham, he critiqued attempts to quantify utilitarianism, instead emphasizing a more qualitative focus. In particular, he pointed out that intellectual and moral pleasures are higher attainments than sheer physical ones, famously saying that "it is better to be a

---

[9] The town of Unity, Vermont is fictitious.

human being dissatisfied than a pig satisfied."[10] Mill's formulation emphasized *happiness* over *pleasure* as the vital component of utility.

Utilitarian thinking is a dominant influence on modern culture. Some well-known sayings exemplify this, such as "The end justifies the means," or "Sometimes you have to do the wrong thing for the right reason." Note that these statements demonstrate how people often struggle with the conflict between principles and the highest good.

The current emphasis in patient-centered health care is to maximize good outcomes, making it very compatible with utilitarian thinking. This trend accords with our ethical intuitions. On a meta-ethical level, it seems reasonable to make "happiness" the measure of utility. After all, happiness seems like an end in itself, a concept that goes back to Aristotle (Kraut 2018). Also, there is something very democratic about utilitarianism, since it makes all moral agents equal; the happiness of one person is no more important than that of another.

On the other hand, even if we reject Bentham's quantifying of utility, all this seems somewhat vague and imprecise. Can we really predict the ultimate outcomes of any particular action? There may be hidden or unknown ramifications that might play a role, and imperfect moral agents might not make the right call. More importantly, some principles indeed matter and should outweigh outcomes.

Classical utilitarianism may fail to protect minorities and their rights adequately (Wilkens 2011). A brief consideration of twentieth century history will reveal several examples, of which we will mention just one. In the southern United States of the 1950s, "Jim Crow" laws segregated different races for the use of restrooms, theaters, and lunch counters, which many have referred to as "the tyranny of the majority"[11] (de Tocqueville and Reeve 1835). These laws deliberately prevented contact between blacks and whites as equals, and therefore reinforced segregation. Sometimes the act that brings the most happiness to the most people is just wrong.

How does utilitarianism help the Hospital Ethics Committee in the fictional town of Unity, Vermont resolve Dr. Davies' question regarding the stockpiling of influenza vaccine to vaccinate children preferentially? First of all, it is essential to note that this is an emergency. Under normal circumstances, the vaccine should be offered to every patient at risk, young and old alike. But under these conditions, with the vaccine in short supply, Dr. Davies' creative solution may seem reasonable.

The Hospital Ethics Committee might plausibly approve the Davies plan since it would bring about the best outcome for the most patients. Stockpiling the vaccine helps to protect the vulnerable pediatric population of Unity while authorities scramble to obtain further vaccine supplies. Note also that Dr. Davies was wise in consulting the Ethics Committee since they could defend her actions if the state's Board of Pharmacy ever questioned her.

---

[10] Mill JS. Utilitarianism (1863). *Utilitarianism, Liberty, Representative Government*. 1859:7–9.
[11] Urofsky MI. Jim Crow Law. *Encyclopedia Brittanica*. www.britannica.com/event/Jim-Crow-law/Homer-Plessy-and-Jim-Crow.

Situation Ethics

We make brief mention here of situation ethics, a relationship-based ethical theory, especially prevalent in the latter part of the twentieth century. Its chief modern proponent was Joseph Fletcher, an Episcopal theologian and bioethicist who later became an atheist. His 1966 book, *Situation Ethics: The New Morality*, was highly influential (Fletcher 1966). Like classical utilitarianism, this theory is mostly consequentialist, only instead of happiness or utility, the ultimate standard is *love*. The working principle is flexibility, considering the circumstances of a particular action to maximize love. One might rightly ask the meta-ethical question: why is love the standard, and not happiness or utility? Situation ethics had a flurry of interest, especially in public education in the U.S. in the 1970s. In practice, situation ethics seems vague and relativistic and is not widely held today (Outka 1998).

Virtue Ethics

Case Study
Nora Simpson is a busy mother of two active toddlers. Trained as a nurse, she now stays at home full-time to care for her little ones. It seems like she is always getting respiratory and intestinal viruses, which she suspects the children bring home from daycare and preschool.

Nora often has coffee with her best friend Becky, also a mother with little children, who seems to be remarkably resistant to such problems. Becky claims that her good health has resulted from taking daily probiotic supplements containing "good bacteria," such as Lactobacillus species, to boost her immune system. As a nurse, Nora is skeptical of such claims, since they are not proven by the FDA.

Uncertain what to do, Nora remembers her local pharmacist, Dr. Anthony Winsome. He has always been kind and engaging with patients who come into his retail pharmacy. His patience and sound judgment have impressed her, and she admires his calm demeanor and moral character.

Nora stops by for a conversation with her pharmacist. After explaining her circumstances, she asks, "Dr. Winsome, if you were in my situation, what would you do?"

Ethics Question
How should Dr. Winsome respond?

Everyone loves a hero. Modern role models with noble character abound, from Mother Teresa to Dr. Martin Luther King. As an ethical theory dating back to Aristotle, *virtue ethics* is all about one's strength of character. Aristotle described *arête* (Gr. "virtue" or "excellence") as the "Golden Mean" between extremes of deficiency and excess (Kraut 2008). An example will help to make this clear. A battlefield solder should not be a reckless, Rambo-like character who risks his own life and that of others. Neither should he be a coward who fails to fight. The "golden mean" might represent simple courage.

Consider another example. What traits make for a virtuous investor, one who handles her own money well? A reasonable investor should not be foolhardy or generous to a fault, giving away money to anyone who asks. On the other hand, we tend not to admire a greedy miser, one who always seeks her own gain by stepping on others. The "golden mean" might be someone with a reputation for thriftiness.

One problem with this approach is the definition of virtue, which may be culture-dependent. Aristotle's idea of *arête* was a bit different than the modern concept (Kraut 2018). Both, however, focus on conceptions of human flourishing, those character traits that build the kind of society we all want. If this seems somewhat vague, it is. Since it focuses on character, virtue ethics does not examine individual actions themselves. Therefore, it does not give guidance for specific ethical problems. Instead, it assumes that the virtuous person "will know what to do." Its value for modern pharmacy practice should be apparent. A good reputation as a clinician takes a long time to establish, but looking to healthcare professionals as moral exemplars is not a bad thing. It builds trust and gives patients confidence in the advice they receive.

How does all this affect Dr. Winsome and his conversation with Nora Simpson? The pharmacist should rightly be pleased that his patient expresses confidence in him. While leaving the final decision up to her, he would not be wrong to share his own opinion concerning probiotics, assuming that he has examined the evidence.

Summary

This chapter has introduced basic definitions in ethics and has organized its theories into the scheme shown below. The next chapter will discuss the various ways we should understand human value and human dignity.

An Organizing Scheme for Ethical Theories

Theories Based on Reason

1. Kantian Ethics
2. Natural Law

Theories Based on Tradition

1. Divine Command Theory (extreme and modified variants)
2. Medical Principlism (Hippocratic Oath + autonomy, discussed later)

Theories Based on Relationships

1. Utilitarianism
2. Situation Ethics (discussed only briefly)
3. Virtue Ethics

Key Terms

- ethics v. morality
- normative ethics
- meta-ethics
- deontological
- consequentialist

Questions for Review and Discussion

1. What makes moral philosophy different from empirical disciplines?
2. Give two implications of the "freedom of action" principle in ethics.
3. What is a potential problem with dividing ethical theories up as deontological or consequentialist?
4. What are the three principal sources of ethical rules, as defined in this chapter?
5. Define and give an example of Kant's First and Second Categorical Imperatives.
6. Name two "fathers" of natural law.
7. What is Euthyphro's Dilemma? Why is it important?
8. What ethical idea was endorsed by William of Ockham? What problem does this create?
9. Name two "fathers" of utilitarianism. Explain how this theory works.
10. What is virtue ethics? Who first proposed this idea?
11. As a review, cite the strengths and weaknesses of each of the major ethical theories.

Further Reading

Marino, Gordon: *Ethics, The Essential Writings.* Modern Library, 2010. A helpful anthology of primary sources.

Shafer-Landau, Russ, *The Fundamentals of Ethics, 2nd Edition,* Oxford University Press, 2011. A scholarly, more philosophical summary.

Wilkens Steve: *Beyond Bumper Sticker Ethics: An Introduction to Theories of Right and Wrong, 2nd Edition.* IVP Academic, 2011. Accessible, popular-level treatment, from a faith-based perspective.

References

Austin MW. Divine command theory. In: Internet encyclopedia of philosophy. 2018. www.iep.utm.edu/divine-c/.

Beauchamp TL, Childress JF. Principles of biomedical ethics. 8th ed. New York: Oxford University Press; 2019.

Brown CM. Thomas Aquinas. Internet encyclopedia of philosophy. 2018. www.iep.utm.edu/aquinas/.

Budziszewski J. What we can't not know: a guide. Revised and expanded ed. San Francisco: Ignatius Press; 2011.

Cameron N. The new medicine: life and death after Hippocrates. Chicago: Bioethics Press; 2001.

References

Campbell CS. Limiting the right to die: moral logic, professional integrity, societal ethos. In: Euthanasia and assisted suicide: global views on choosing to end life, 191. 2017.

Connolly PJ. John Locke. Internet encyclopedia of philosophy. 2020. www.iep.utm.edu/locke/.

de Tocqueville A, Reeve H. Democracy in America. London: Saunders and Otley; 1835.

Duigna B. Naturalistic fallacy. In: Encyclopædia Britannica. 2012. www.britannica.com/topic/naturalistic-fallacy.

Edelstein L. The Hippocratic oath, text, translation and interpretation. Baltimore: Johns Hopkins Press; 1943.

Feinberg JS, Feinberg PD. Ethics for a brave new world. 2nd ed. Wheaton: Crossway; 2010.

Fieser J. Ethics. In: Internet encyclopedia of philosophy. 2018. www.iep.utm.edu/ethics/.

Fletcher JF. Situation ethics; the new morality. Philadelphia: Westminster Press; 1966.

Foot P. The problem of abortion and the doctrine of double effect. Oxf. Rev. 1967;5:5.

Foot P. The problem of abortion and the doctrine of the double effect. Oxford: Oxford University Press; 2002.

Johnson R, Cureton A. Kant's Moral Philosophy. In: Stanford encyclopedia of philosophy; 2016. https://plato.stanford.edu/entries/kant-moral/.

Jonsen AR, Siegler M, Winslade WJ. Clinical ethics: a practical approach to ethical decisions in clinical medicine. 8th ed. New York: McGraw-Hill Education; 2015.

Kraut R. The Blackwell guide to Aristotle's Nicomachean ethics. New York: Wiley; 2008.

Kraut R. Aristotle's ethics. In: Stanford encyclopedia of philosophy. 2018. https://plato.stanford.edu/entries/aristotle-ethics/.

Macleod C. John Stuart Mill. Stanford encyclopedia of philosophy. 2016. https://plato.stanford.edu/entries/mill/.

Murphy M. The natural law tradition in ethics. In: The Stanford encyclopedia of philosophy. 2011. https://plato.stanford.edu/archives/win2011/entries/natural-law-ethics/.

O'Neill O. A simplified account of Kant's ethics. In: Cahn SM, editor. Exploring ethics: an introductory anthology. New York: Oxford University Press; 2009.

Outka G. Situation ethics. In: Routledge encyclopedia of philosophy. 1998. www.rep.routledge.com/articles/thematic/situation-ethics/v-1

Shafer-Landau R. The fundamentals of ethics. 4th ed. New York: Oxford University Press; 2018.

Sweet W. Jeremy Bentham. Internet encyclopedia of philosophy. 2018. www.iep.utm.edu/bentham/.

Veatch RM. Medical codes and oaths. In: Encyclopedia of bioethics. 2nd ed. New York: Macmillan; 1995.

Wilkens S. Beyond bumper sticker ethics: an introduction to theories of right and wrong. 2nd ed. Downers Grove: IVP Academic; 2011.

Chapter 3
Human Value and Human Dignity

Having introduced the major ethical theories, we now turn to a discussion of human value. Why should pharmacists and pharmacy students care about this historical debate? The reason is simple: the pharmacy profession is *relational*. Recent surveys have consistently shown that students enter the pharmacy profession for two main reasons: they love the sciences, and they want to work with people (Hanna et al. 2016; Willis et al. 2006; Keshishian 2010; Capstick et al. 2007). To a substantial degree, *people* are at the heart of the profession; relating to others and helping them with their healthcare needs is a primary motivator. In this chapter, we will discuss the importance of clear and accurate thinking about the value of people. We will present two possible ways of defining such value, as well as two competing definitions of human dignity. This approach will help the reader better understand how the framing of such ideas affects historical ethical controversies, especially at the beginning of life.

Human Personhood

The term *personhood* refers to something more than mere biological life. To be a *person* means to be "a member of the moral community" (Bagnoli 2007). Another word for a person is a *moral agent*, one who has moral rights and duties. The term implies value, dignity, and respect. The ethical landscape is dominated by two major theories of human personhood, each of which has ancient roots. These two ideas are *empirical functionalism* and *ontological personalism*.

Empirical functionalism is the view that a set of functions or abilities defines human personhood. The term "empirical" comes out of the scientific method and refers to directly measurable traits (Merriam-Webster.com 2018). Although its roots are ancient, the classical modern expression of functionalism was articulated by bioethicist Joseph Fletcher, who in 1972 outlined twenty criteria for human personhood (which he called "humanhood"). These criteria included such hallmarks as minimum intelligence, self-awareness, a sense of time, and the capacity to relate to others (Fletcher 1972). In response, other philosophers weighed in, one emphasizing self-awareness (Tooley 1972), and another focusing on "relational potential," based on the ability to interact socially with others (McCormick 1974). From the feedback of these and other writers, Fletcher decided that the minimum essential requirement for moral status was the full functioning of the human neocortex (Fletcher 1974). Neocortical functions are those "higher brain" processes of the cerebral cortex necessary for active consciousness and volition. This idea should be contrasted with whole-brain functioning, which includes activities of the brainstem as well as the cortex.

What are the moral implications of this view? Much of the debate centers around human development in the womb. Since Fletcher, functionalists have used the above criteria to claim that the unborn have no moral status since they lack rationality or self-awareness. However, by this criterion, one could also argue that adults lack self-awareness when they are asleep or under anesthesia, yet no one questions their moral status during such moments. One way to circumvent this objection is to use the idea that only "continuing selves" have personhood, which includes both self-awareness and a sense of the future (Tooley 1983). In practical terms, for example, these ideas have provided a primary moral justification for induced abortion at any stage of pregnancy.

Michael Tooley and more recently, the Princeton philosophy professor Peter Singer, have both advocated the next logical step: infanticide.[1] If the fetus has no right to personhood because it is not yet self-aware, then neither does the unwanted newborn with various congenital disabilities. According to Singer, "Infanticide before the onset of self-awareness . . . cannot threaten anyone who is in a position to worry about it" (Kuhse and Singer 1985).

Some authors find such a position objectionable and argue that there is a difference between beings with a *potential* capacity for rationality and those with a *developed* capacity. Though the former are not persons (on this view), both are entitled to a right to life, with that right increasing with greater and greater development of potential (Wennberg 1985). Others compare personhood to a process: "When can we say that the fetus is a human being rather than a human becoming? Surely only when its metamorphic process is complete" (Becker 1988). These ideas would imply, therefore, that there are *degrees* of personhood, i.e., that some human beings have more of it than others. On this view, there can be such an entity as a *human non-person*.

Functionalist perspectives have the advantage of being concrete. Cognitive functions can be measured empirically, and they accord with many intuitive ideas about moral status. For example, surveys have shown that more than half of all Americans

---

[1] Veith GE. Professor Death: Princeton hires a pro-death ethicist vilified by the rest of the world. *World Magazine,* Published 8/8/1998. https://world.wng.org/1998/08/professor_death.

support the legality of abortion; however, that support decreases significantly with progressive stages of fetal development and is very low in the third trimester.[2]

On the other hand, functionalism may seem to go against commonly held values of fairness and equality. To claim that some human beings do not have moral worth because they lack cognitive abilities may seem to violate the principles of a just society. However, some writers have taken the ideas of Tooley and Singer even further. For example, Giubilini and Minerva have coined a new phrase: "after-birth abortion." They mean for this to refer to the taking of a newborn baby's life, even if the child is perfectly healthy, if any manner of external factors make having a new baby difficult for the parents. The abstract of their article in the *Journal of Medical Ethics* puts it this way:

> Abortion is largely accepted even for reasons that do not have anything to do with the fetus' health. By showing that (1) both fetuses and newborns do not have the same moral status as actual persons, (2) the fact that both are potential persons is morally irrelevant and (3) adoption is not always in the best interest of actual people, the authors argue that what we call 'after-birth abortion' (killing a newborn) should be permissible in all the cases where abortion is, including cases where the newborn is not disabled (Giubilini and Minerva 2012).

In coining the phrase 'after-birth abortion,' the authors attempted to de-stigmatize what otherwise would be called infanticide. This logical extension of functionalist views was too extreme for many, and the clinical ethics community has widely condemned the published article.

Philosopher Peter Singer's influence on these matters also stems from his claim that functional personhood should extend to other mammalian species than *Homo sapiens*. Attempts to privilege human beings as persons commit an offense that Singer calls *speciesism*. Here is how he expresses the idea in his book, *Practical Ethics*: "To give preference to the life of a being simply because that being is a member of our species would put us in the same position as racists who give preference to those who are members of their race" (Singer 2011). On his view, the unborn fetus or the cognitively-impaired adult has a lower moral status than that of an adult chimpanzee or ape. Singer holds that any other conclusion derives from religious ideas, which as an atheist he rejects.

In summary, empirical functionalism defines human personhood by a set of functions or abilities, usually couched in terms of cerebral development or self awareness. As cognitive abilities increase, so does the degree of personhood.

The second major theory of human personhood is *ontological personalism*, which states that all human beings are human persons, by mere virtue of membership in the species *Homo sapiens*. On this view, the intrinsic quality of personhood begins at conception and is present throughout life. The word 'ontological' here is a philosophical term that means "relating to or based upon being or existence".[3] This emphasizes that the view derives from reason and logic alone; i.e., human beings

---

[2] Gallup. Abortion. https://news.gallup.com/poll/1576/abortion.aspx. Published 2019. Accessed 2/15/2021.

[3] Ontological. *Miriam-Webster Online*. 2020. www.merriam-webster.com/dictionary/ontological.

are not potential persons or "becoming" persons; they are persons by their very nature. According to ontological personalism, there is no such thing as a potential person or a human non-person (Williams and Bengtsson 2018; O'Mathuna 1996; Palazzani 2017).

Herein lies the tension between the two ideas of personhood. As society has become more secular, an automatic, transcendental reverence for life has been replaced with a more reductionist view. The empirical functionalist concept of personhood holds that a human being is merely a collection of parts and functions, or a *property-thing*. Put together enough chemical molecules in the right way, and you have a human being; put another set of elements together, and you have a 1957 Chrysler. Philosophically, it makes no difference.

Ontological personalism, on the other hand, is based on the premise that a human being is a *substance*. A substance is a distinct unity of essence that exists *ontologically prior* to any of its parts. This traditional concept dates back to Aristotle and Thomas Aquinas, but modern philosophers such as J. P. Moreland and Alvin Plantinga also adhere to this view (Moreland and Craig 2009; Moreland and Rae 2000; Robinson 2018). To make the complex idea of substance more clear, we will focus on two of its implications: the parts v. whole distinction, and continuity.

Let us expand on our earlier illustration of a classic automobile. Consider a nicely restored 1957 Chrysler – many of its original parts have rusted away and have been replaced, so that this vintage car is a collection of old and new. Although many will refer to it as the *same* car as when it was new, intuition tells us that this is not the case. In fact, as stated earlier, remove the wheels, the motor, the seats, and the body, and the result is no longer a 1957 Chrysler; it is not even a car. To go still further, imagine adding other parts to the original chassis, such that the result is a 1972 Volkswagen Beetle. There was no continuity of essence between the two vehicles; each is nothing more than a collection of parts.

Try to do the same kind of thought experiment on a human being. Remove an arm or a leg from John Doe, and he remains a person, in fact, the *same* person. You can amputate all of John's extremities and even remove many internal organs; as long as he remains alive, his substance will never change. You can even "add new parts," by transplanting organs from other persons, yet John Doe will never become James Smith; his component parts do not define his substance. He will always remain the same person, and the whole is *more than* his parts.

The second implication of substance is the argument from continuity. The cells of the human body are continually being replaced. As nutrients are taken in and waste products given off, new chemical molecules enter and leave daily. The outer skin is completely replaced every four weeks. The lining of the gastrointestinal tract is replaced even more rapidly, every week or so (Tortora and Derrickson 2018). It is reasonable to claim that all of the chemical parts of the body are completely replaced, say, every few years.

Nonetheless, an individual, as a substance, has continuity from one moment to the next. She is the same person as she was one week ago, one year ago, or ten years ago. She has memories that give her continuity with her present state. She relates to her childhood; she can give the date of her birth. Even if she lacks such memories because of disease or injury, she has a continuing self that is identical to her earlier self.

Strict naturalism has considerable difficulty here. To hold to a property-thing view of persons is to deny the commonsense understanding of personal continuity, with a host of attendant problems for law and morality. In applying these ideas to moral debates, let us briefly consider the unborn. We can use intuition and continuity arguments to argue persuasively for personhood. Using common sense, there is no *prima facie* (at first impression, or self-evident) reason to assume that a baby changes its essential nature by virtue of geography (namely, in the womb or out of it). And there is no *prima facie* reason not to extend such humanity further back in time. In fact, the continuity argument argues for the personhood of the fetus back to the moment that it first became a substance, i.e., the moment of conception.

The continuity argument is also consistent with religious arguments articulated by the great monotheistic faith traditions. Jewish and Christian scriptures consistently teach the value of the unborn. Psalm 139 is one example: "For you created my inmost being; you knit me together in my mother's womb" (v. 13, NIV).

It is worth noting that the Jewish and Christian scriptures also ground human value in the idea that God created humankind in his image. This approach allows for explorations of the way human beings resemble the Divine (rationality, volition, social nature, etc.) while avoiding the limitations of a strictly functional definition. On this view, the image of God is *intrinsic* to the nature of persons. Thus, Jewish and Christian scriptures teach the value of human beings generally and occasionally refer to their value in the womb. It would, however, be inadvisable to claim that such pre-modern writings specify when human value begins. Nonetheless, intuition and philosophy have led many to conclude that such a valuation starts at conception.

Beyond the abortion debate, these competing definitions of human value inform many other areas of biomedical ethics. At the other end of the spectrum, these ideas also affect elderly patients with cognitive disabilities. Think, for instance, of an elderly nursing home resident with Alzheimer's disease. How her family thinks about her personhood may subtly influence whether or not to insert a feeding tube if she is unable to eat for herself. Similarly, views of human personhood significantly influence discussions of medical futility, palliative care, hospice, euthanasia, and assisted suicide, to name just a few issues.

Summary of Personhood Concepts

Person (def.): moral agent, a member of the moral community.
Empirical functionalism: our functions or abilities determine personhood.
Ontological personalism: all human beings are human persons, by virtue of membership in our species.

Human Dignity

What is dignity? What does it mean? According to some recent scholars, not much. Consider medical ethicist Ruth Macklin, who wrote in 2003 that dignity is a hopelessly vague term. She writes:

> In the absence of criteria that can enable us to know just when dignity is violated, the concept remains hopelessly vague ... Dignity is a useless concept in medical ethics and can be eliminated without any loss of content (Macklin 2003).

Expressed even more strongly, psychologist Steven Pinker claims that dignity is a "stupid" term, at best a "squishy, subjective notion, hardly up to the heavyweight moral demands assigned to it" (Pinker 2008). Both Macklin and Pinker seem to be saying that appeals to dignity, at their base, are nothing more than appeals to respect for personal autonomy. Of course, autonomy is an essential concept under medical principlism, as we will discuss in more detail in a later chapter. But does dignity have a place in medical ethics discussions? If so, how should we define it, and what role should it play?

One helpful approach to this question comes from the Universal Declaration of Human Rights, ratified by the United Nations in 1948 (Universal Declaration of Human Rights 1948). The authors wanted a basis for giving people inalienable rights, yet they wanted to do this without appealing to religion, philosophy, or politics. Philosopher Glenn Hughes summed up their efforts in this way:

> [O]n the one hand, they were all universalists, in the sense "that they believed that human nature was everywhere the same and that the processes of experiencing, understanding, and judging were capable of leading everyone to basic truths" about the human condition, such as the truth that humans have an inherent dignity or worth; while at the same time, they were all pluralists, in holding that such basic truths may receive authentic, more or less equivalent articulation in different linguistic manners in historically diverse cultures (Hughes 2011).

Therefore they decided that the rights outlined in the Universal Declaration stem from intrinsic dignity, a "founding explanatory principle that was both universal and pluralistic." According to Hughes, some see this ambiguous definition of dignity as a weakness of the declaration. He, however, sees it as a strength. Dignity becomes open-ended and allows for the development of four characteristics: liberty, responsibility, irreplaceability, and vulnerability (Hughes 2011).

Note that this definition of dignity seems to relate closely to the ontological understanding of personhood, as discussed earlier. However, deriving an intrinsic idea of dignity from a pluralistic frame, without recourse to religion, opens up to a broad range of criticisms. What is the *source* of such inherent dignity? How can we claim that it applies to all human beings?

Viewed in a slightly different way, dignity seems to be achievable and variable, depending on human conditions. The anencephalic infant (lacking a functioning cerebral cortex), the comatose adult patient, the elderly woman with intractable cancer pain: each of these seems to *lack* dignity in some way analogous to a functional understanding of personhood. As an example application, this idea is at the heart of "death with dignity" laws permitting physician-assisted suicide in various U.S. states and throughout Canada.

So depending on the context, the concept of human dignity may be applied to every human being, as in the statement, "All human beings have inherent dignity, regardless of their age or life circumstances." Or alternately, some would claim that people may *lose* their dignity based on painful, uncomfortable, or embarrassing clinical circumstances and that this should be a guide for ethical decision-making. No matter which view one chooses, the concept of dignity, though vague, remains essential and applicable to the modern clinical context. The following example will illustrate.

>Case Study:
>Jonathan Franken is a 26-year-old man who presented to a family-owned retail pharmacy with a prescription for 30 Percocet 5/325 tablets. The directions stated: take 1 or 2 tablets every 6 h as needed for pain. A prominent local general surgeon had signed the prescription.
>Christopher Macklin was the pharmacist on duty, who noticed that Jonathan had tattoos and track marks on both arms. He sported a ponytail and was wearing a greasy-looking biker T-shirt. Dr. Macklin searched his state's online prescription monitoring program and noticed that Jonathan had obtained controlled substances at three different pharmacies in the past two months.
>The pharmacist then challenged the patient, loudly saying, "What's this for? Are you doctor shopping? That won't work here!"
>Jonathan looked behind him at the three other patients waiting in line and turned red with embarrassment. He grabbed the prescription and rushed out of the building. Later that day, he visited a nearby chain pharmacy, where he received his medication. He returned home, glad that he could get some relief for his post-operative pain from the groin hernia surgery he had undergone the day before.

What went wrong in this situation? Dr. Macklin had made certain judgments about the patient, as many of us are inclined to do as humans and professionals. However, his response failed to respect Jonathan Franken's dignity, and also violated HIPAA privacy laws, causing confusion and concern among the other patients present.

To be clear, Dr. Macklin had every legal right, in fact, a duty, to question the purpose of the opioid prescription, in light of a possible history of abuse. This concern should have led to a phone call to the prescribing physician to verify that the prescription was legitimate. Afterward, the pharmacist should have taken the patient to a private area, where he could have explained his concerns and received more information. The result would have been a therapeutic benefit to the patient and the preservation of their professional relationship. At the heart of this problem was a failure of Dr. Macklin to respect personal dignity.

In summary, human value and human dignity have a significant bearing on the ethics of any clinical interaction. This chapter has presented two alternative philosophical ways of defining human personhood: empirical functionalism and ontological personalism. We have also examined two contrasting views of human dignity, one intrinsic and immutable, the other acquired and based on life events. If these competing formulations of personhood and dignity have appeared abstract and confusing, it is because these terms are often thrown around loosely in ethics discussions and in everyday language. The perceptive reader will be aware of these distinctions, and the context will usually reveal which idea is intended by various healthcare professionals.

Key Terms

- person (or moral agent)
- empirical functionalism
- ontological personalism
- speciesism
- property-thing
- substance
- *prima facie* argument
- human dignity

Questions for Review and Discussion

1. Name the two major philosophical theories of human personhood.
2. Define a "substance" and contrast this with a property-thing.
3. Discuss how the two competing views of human personhood may influence clinical decision-making in a severely deformed newborn infant.
4. Discuss the concept of human dignity. Why is it controversial?

Further Reading

Kass, Leon. *Being Human: Core Readings in the Humanities*. W.W. Norton, 2004. An exploration of human value through the lens of various genres of classic literature.

Meilaender, Gilbert. *Neither Beast nor God: The Dignity of the Human Person*. Encounter Books, 2009. A philosophical discussion of the concept of dignity.

Pellegrino, Edmund, et al. *Human Dignity and Bioethics*. University of Notre Dame Press, 2009. An edited collection of essays on dignity and human value from the President's Council on Bioethics.

References

Bagnoli C. Respect and membership in the moral community. Ethical Theory Moral Pract. 2007;10(2):113–28.

Becker L. Human being: the boundaries of the concept. In: Goodman MF, editor. What is a Person? Clifton: Humana Press; 1988.

Capstick S, Green JA, Beresford R. Choosing a course of study and career in pharmacy – student attitudes and intentions across three years at a New Zealand School of Pharmacy. Pharm. Educ. 2007;7:359–73.

Fletcher JF. Indicators of humanhood: a tentative profile of man. Hast. Cent. Rep. 1972;2(5):1–4.

Fletcher JF. Four indicators of humanhood – the enquiry matures. Hast. Cent. Rep. 1974;4(6):4–7.

References

Giubilini A, Minerva F. After-birth abortion: why should the baby live? J. Med. Ethics. 2012; https://doi.org/10.1136/medethics-2011-100411.

Hanna L-A, Askin F, Hall M. First-year pharmacy students' views on their chosen professional career. Am. J. Pharm. Educ. 2016;80(9):150.

Hughes G. The concept of dignity in the universal declaration of human rights. J. Relig. Ethics. 2011;39(1):1–24.

Keshishian F. Factors influencing pharmacy students' choice of major and its relationship to anticipatory socialization. Am. J. Pharm. Educ. 2010;74(4):75.

Kuhse H, Singer P. What's wrong with the sanctity of life doctrine? In: Should the baby live: the problem of handicapped infants. Oxford: Oxford University Press; 1985. p. 118–39.

Macklin R. Dignity is a useless concept. Br. Med. J. 2003;327:1419–20.

McCormick RA. To save or let die: the dilemma of modern medicine. JAMA. 1974;229:172–6.

Merriam-Webster.com. Empirical. 2018. www.merriam-webster.com/dictionary/empirical.

Moreland JP, Craig WL. Philosophical foundations for a christian worldview. Downers Grove: InterVarsity Press; 2009.

Moreland JP, Rae SB. Body and soul. Downers Grove: InterVarsity Press; 2000.

O'Mathuna D. Medical ethics and what it means to be human. Ir. Bible Sch. J. 1996:12–9.

Palazzani L. Person and human being in bioethics and biolaw. In: Legal personhood: animals, artificial intelligence and the unborn. Cham: Springer; 2017. p. 105–12.

Pinker S. The stupidity of dignity. The New Republic. 2008;28(05.2008).

Robinson H. Substance. In: Stanford encyclopedia of philosophy. 2018. https://plato.stanford.edu/entries/substance/.

Singer P. Practical ethics. 3rd ed. New York: Cambridge University Press; 2011.

Tooley M. Abortion and infanticide. Philos. Public Aff. 1972:37–65.

Tooley M. Abortion and infanticide. Oxford: Clarendon Press; 1983.

Tortora GJ, Derrickson B. Principles of anatomy & physiology. 15th ed. Hoboken: Wiley; 2018.

Universal Declaration of Human Rights. UN General Assembly. 1948;302(2). www.un.org/en/universal-declaration-human-rights/.

Wennberg RN. Life in the balance. Grand Rapids: William B. Eerdmans Publishing; 1985.

Williams TD, Bengtsson JO. Personalism. In: Stanford encyclopedia of philosophy. 2018. https://plato.stanford.edu/entries/personalism/.

Willis S, Shann P, Hassell K. Who will tomorrow's pharmacists be and why did they study pharmacy? Pharm. J. 2006;277(7410):107–8.

Chapter 4
Clinical Ethics in Historical Context, Part I

Medical Principlism and the Hippocratic Tradition

No historical account of healthcare ethics can fail to mention the enormous influence of Hippocrates on the practice of modern medicine. The ancient Hippocratic Oath, written either by the great doctor himself or his followers, is relatively short. Still, it contains three of the four core ethical tenants found in medical principlism, the modern standard for healthcare ethics. The principles are beneficence, nonmaleficence, distributive justice, and autonomy. We briefly mentioned these in Chap. 2, but we will now define and discuss each in turn (Beauchamp and Childress 2019).

We do not know much about Hippocrates himself. He was born on the Greek island of Cos (c. 460 BCE), where he founded a school of medicine that was distinct from the culture of his day. One of his influences was Aesculapius, a demi-god in Greek mythology. Like his father Apollo, Aesculapius had great healing powers. The cult of Aesculapius believed that diseases were due to the displeasure of various gods. They taught that healing could be achieved by interpreting dreams and by the laying-on of non-poisonous snakes (Savel and Munro 2014). Probably also influenced by the philosopher Pythagoras, Hippocrates began to depart from these superstitious views in favor of a more rational explanation of illness. In particular, he adopted the Pythagorean idea that diseases arose by an imbalance of four humors (blood, yellow bile, black bile, and phlegm).

Though the humoral theory of disease is now outdated, it was, at the very least, a testable hypothesis, long before the empirical methods of science. A more enduring legacy came about through the fact that Hippocrates and his followers were the first to employ case histories. They also first recognized that similar symptoms were due to similar diseases (Fig. 4.1) ('Hippocrates' 2014).

Though his medical diagnoses and treatments were influential and helpful in their day, it is in his ethics that Hippocrates has had his most significant impact. His respect for humanity and sensitivity to the needs of individuals set him apart from

Fig. 4.1 Hippocrates. (Image in public domain: 1881 young persons' Cyclopedia of persons and places)

all his contemporaries. Anthropologist Margaret Mead has pointed out that before the Hippocratic era, the roles of physician and sorcerer were indistinguishable (Bulger and Barbato 2000). One author expressed it this way: "Before Hippocrates, the hapless patient could never be certain when he hired a doctor for some white magic that one of his enemies had not paid to dispatch him with black magic" (Del Guercio 1977). The unique commitment of Hippocratic physicians to the well-being of their patients made them distinct from other practitioners of the healing arts.

The opening paragraph of the Hippocratic Oath is instructive:

> I swear by Apollo Physician, and Aesculapius, and Hygeia, and Panacea, and all the gods and goddesses, making them my witnesses, that I will fulfill according to my ability and judgment, this Oath and this covenant...[1]

First of all, note the solemn and formal vertical dimension: this is a pledge made to pagan deities. This idea is amplified by the phrase "this oath and this covenant," implying a sacred duty that goes beyond a mere contract. Though the pagan gods and goddesses mentioned are no longer highly regarded, the ideas contained in the Oath are compatible with the great monotheistic religious traditions of the Western world, which helps explain their endurance. This emphasis has also contributed to the idea of healthcare as a spiritual "calling" (Fig. 4.2).

After a pledge to honor one's teachers, the Oath continues:

> I will apply [treatment] for the benefit of the sick according to my ability and judgment; I will keep them from harm and injustice. I will neither give a deadly drug to anybody if asked for it, nor will I make a suggestion to this effect. Similarly I will not give to a woman an abortive remedy (Edelstein 1943).

---

[1] Translation from the Greek by Ludwig Edelstein: The Hippocratic oath, text, translation and interpretation. Baltimore,: The Johns Hopkins press; 1943.

Medical Principlism and the Hippocratic Tradition

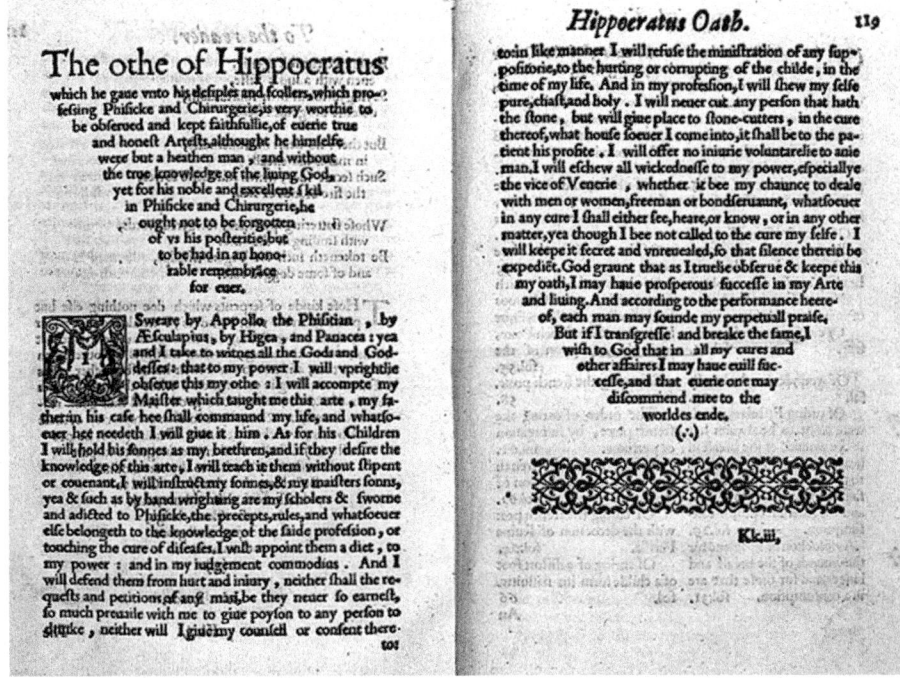

Fig. 4.2 Pages from a sixteenth-century medical text. (F. Arcaeus, pages from 'A most excellent and compendious method of curing woundes in the head...' Credit: Wellcome Collection. Attribution 4.0 International (CC BY 4.0))

Today, we summarize the first part of all this with the term *beneficence*. The word comes from the Latin *beneficentia*, meaning "the quality or state of doing or producing good" ('Beneficence' 2020). The principle states that healthcare professionals should always act in the best interests of their patients. If this seems too obvious, remember the history of the Hippocratic tradition, that the doctor as a healer was not always a given. The phrase, "according to my ability and judgment," implies a high degree of professionalism and expertise.

> **Beneficence**
> Healthcare professionals should always act in the best interests of their patients.

Closely allied with beneficence is the term *non-maleficence*, which derives from the words of the Oath, "I will keep them from harm and injustice." Many readers will recall the famous phrase, "First of all, do no harm." Surprisingly, Hippocrates never said this exact phrase, nor its Latin equivalent, *primum non nocere*, which derives from a much later era (Sokol 2013). Yet the idea of avoiding harm to patients was a fundamental principle of Hippocratic practice, as he states in his *Epidemics*, "The physician must … have two special objects in view with regard to disease, namely, to do good or to do no harm" ('Hippocrates' 2009).

Avoiding harm is such an essential part of Hippocratic ethics that the great doctor gives two specific illustrations of it: "I will neither give a deadly drug to anybody if asked for it, nor will I make a suggestion to this effect. Similarly I will not give to a woman an abortive remedy." The first of these prohibitions was a response to the Greek culture of that era, in which suicide was a common solution to serious illness (Gourevitch 1984). The very specificity of this statement, which forbids even a mention of the possibility of helping with suicide, demonstrates an absolute commitment to a particular form of medical professionalism: the doctor knows best (Cameron 2001). Likewise, the physician is enjoined from ever performing an abortion, usually by pessary. Once again, induced abortion, though difficult and prone to complications, was common in ancient Greece (Edelstein et al. 1967).

Non-maleficence
Do no Harm.

There is a curious linkage of these ideas in the Oath. Why does Hippocrates list two specific practices, common in his day, as forbidden? The connecting word "similarly" shows that the two examples are part of the same overall professional commitment. Leon Kass suggests that the Hippocratic practitioner "refuses to participate directly in ending a life whether in the fullness of days or on the way to birth" (Guinan 2011). Medical historian Ludwig Edelstein traces these unique standards to the influence of Pythagoras, who, as we have already seen, was an important influence on Hippocrates (Edelstein 1943). In modern medicine, by contrast, elective abortion is legal and common, and there is a growing trend in favor of legalization of assisted suicide. A full discussion of these topics will take place in later chapters, but it should be clear that both practices run counter to the original wording of the Oath.

Our next primary ethical principle comes from the following section of the Hippocratic Oath:

Whatever houses I may visit, I will come for the benefit of the sick, remaining free of all intentional injustice, of all mischief and in particular of sexual relations with both female and male persons, be they free or slaves (Edelstein 1943).

The cultural context of this is instructive. In the so-called classical era of Greece, covering the period between the Persian Wars and the death of Alexander the Great (roughly 510 to 323 BCE), it was a man's world, and women had few rights. Also, slavery was common during this time and well-accepted (Edelstein et al. 1967). So it seems especially curious that Hippocrates would "level the playing field" for everyone, and even protect such marginalized groups. In our modern day, where equal rights and tolerance are the norm, it may be hard to imagine just how insightful and counter-cultural Hippocrates was in this regard. True medical professionalism protects every person, no matter his or her value in the eyes of the rest of society. We now call this principle *distributive justice*.

Distributive justice refers to the fair and equitable distribution of the benefits of healthcare (Beauchamp and Childress 2019). In modern society, it means that there can be no barriers for patients to access clinical treatments, regardless of age, gender, social class, ethnicity, sexual orientation, ability to pay, religion, handicap, or any other medically non-relevant trait. Each patient must be treated in an honorable and respectful manner. As we discussed in Chap. 3, a proper understanding of the personhood and dignity of each patient is a prerequisite to the clinical encounter. The Hippocratic insight on justice has many nuances, which will repeatedly arise in subsequent chapters.

> **Distributive Justice**
> The fair and equitable distribution of the benefits of healthcare.

Our final comment from the Oath will consider this section:

> What I may see or hear in the course of the treatment or even outside of the treatment in regard to the life of men, which on no account one must spread abroad, I will keep to myself, holding such things shameful to be spoken about (Edelstein 1943).

This statement, of course, refers to the principle of *confidentiality*, the idea of medical privacy, and the keeping of secrets by healthcare professionals. As important as it is, most writers list confidentiality as a subset of non-maleficence, and we will follow this practice. It is easy to see how the disclosure of sensitive private information to another party without the patient's permission can cause grave harm (Beauchamp and Childress 2019). Confidentiality is now enshrined in a complex federal law called the Health Insurance Portability and Accountability Act, or HIPAA.[2]

So far, we have examined three of the four major principles of modern medical principlism: beneficence, non-maleficence, and distributive justice, and have shown their ancient roots in the Hippocratic Oath. Our final major principle comes from a much later period in history, the Enlightenment. The Enlightenment, also called the Age of Reason, was a movement in Europe between 1685 and 1815, where science and philosophy threw off their dependence on religion and the Church. One of its principal architects was a German scholar we have already met (Chap. 2), Immanuel Kant. He identified enlightenment "with the process of undertaking to think for oneself, to employ and rely on one's own intellectual capacities in determining what to believe and how to act" (Bristow 2017). It should, therefore, be no surprise that he is the father of the ethical principle called *autonomy*.

---

[2] Health Information Privacy. U.S. Department of Health and Human Resources Web site. www.hhs.gov/hipaa/index.html. Published 2020. Accessed 4/9/2020.

Autonomy comes from the Greek for "self-rule" or "self-law" ('Autonomy' 2020). Respect for autonomy means allowing patients and their surrogates to make their own decisions based on their personal values. Beauchamp and Childress describe this as both a negative and a positive obligation. The negative sense means that a patient must be free of controlling restraints by others in their decision-making. The positive aspect requires the full disclosure of important information, as well as other actions that help to foster independence of thought and decisions on the part of patients and families (Beauchamp and Childress 2019).

> **Autonomy**
> Patients and their surrogates can make their own decisions based on personal values.

Informed Consent and Decision-Making Capacity

Modern medical principlism articulates the four key ideas of beneficence, non-maleficence, distributive justice, and autonomy. It is easy to see that these principles would necessitate a clear understanding of the risks and benefits of a proposed medical intervention before the prescription is written or the procedure performed. This is the idea of *informed consent*.

Informed consent derives primarily from personal autonomy, which means "acknowledging the moral right of every competent individual to choose and follow his or her own plan of life and actions" (Jonsen et al. 2015). To be clear, this is mostly a phenomenon of the twentieth century and later. The professionalism that, at best, protected patients with "the doctor knows best" was also a form of paternalism that failed to give patients any real voice in medical decision-making.

It may surprise readers that even basic consent is a relatively recent doctrine. In 1914, the New York Court of Appeals rendered a judgment on behalf of Mary Schloendorff after doctors performed a surgical procedure on her against her will (Faden et al. 1986; Requarth 2015). In fact, the phrase "informed consent" did not become a part of the law and standard medical practice until the 1950s and 1960s (Beauchamp 2011).

What are the elements of informed consent? To a large extent, consent is contractual: the clinician explains, and the patient agrees. There are three elements to this principle. First of all, the 'informed' part implies full disclosure, and answers the question, "What should the clinician tell the patient to help him or her make a good decision about treatment?" This duty usually means a complete description of the patient's current clinical status, an explanation of possible treatments, along with the risks and benefits of each, and a recommendation based on the clinician's best professional judgment (Jonsen et al. 2015).

Next, informed consent entails an assessment of the patient's comprehension. The healthcare team should endeavor to overcome barriers to an adequate understanding of the information offered. It may be surprising how often the patient truly does not understand the information provided. Some of the reasons may include language barriers, educational level, low health literacy level, patient denial, embarrassment, fear, or subtle coercion. That is why it is valuable to have information presented by multiple team members, not just the physician, pharmacist, or nurse.

Finally, the consent of the patient must be documented on the medical record. This step legally protects both the patient and the healthcare professional. This element is more than just having a signature on file; truly informed consent requires multiple forms of documentation and is an example of shared decision-making (Jonsen et al. 2015).

> **Obtaining Informed Consent – Three Elements**
> - Fully disclose the clinical facts
> - Assure patient comprehension
> - Document the patient's understanding and consent

Who has the power to give consent for medical treatments? After all, informed consent suggests that the patient is fully conscious and mentally and emotionally capable of agreeing to the treatment plan recommended. But a medical crisis may make it impossible for a patient to give consent concerning a clinical intervention, and the situation may be urgent, requiring a quick decision. This raises the issue of *decision-making capacity*. Some patients, either temporarily or on a long-term basis, are unable to give consent to treatments voluntarily. The problem may be severe pain, anxiety, or physical impairment, or their inability may result from a cognitive defect (e.g., a stroke or head injury). Such individuals are said to have lost decision-making capacity (or decisional capacity) (Jonsen et al. 2015; Ganzini et al. 2005).

For the overwhelming majority of healthcare interactions, the clinician assumes that the patient has decision-making capacity; this is not usually in doubt. For some, however, their disease process may interfere with cognitive functioning, raising concerns about their autonomous ability to give consent. One common misconception is that decision-making capacity and *competency* are the same, but this is not true. Decisional capacity is a clinical judgment, usually made by a physician. In contrast, competency is a legal determination made by courts (Ganzini et al. 2005). The difference between capacity and competence is a frequent source of confusion for clinicians.

As an example, consider an 88-year-old woman who recently suffered a stroke, causing her to sleep excessively and making her hard to arouse. Clinicians have recommended placement of a feeding tube to maintain her nutrition, but the woman seems to resist this idea when she is awake enough to respond. What is the next step? The healthcare team should ask for a neurology consult to assess her decision-making capacity and determine if her refusal has a rational basis. Note that family members may ultimately seek a court determination of the woman's competency to manage her financial assets and living situation over the long term. Still, this legal recourse is beyond the purview of her present healthcare interaction.

With this background in mind, there are three possible levels of consent. The highest and preferred level is informed consent by a patient with decision-making capacity. If the patient is a child under the age of 18, an adult with a mental disability, or otherwise lacks decision-making capacity due to a medical issue, the next level is a *substituted judgment* based on decision-making by a surrogate. Who can serve in this role? For most jurisdictions in the U.S., the default surrogate is a family member, with the hierarchy in this order: the patient's spouse, adult child, a parent, and an adult sibling. Not all states require strict adherence to this sequence (Pope 2012).

Some patients have a pre-existing legal representative (this may or may not be a family member) who has a *durable power of attorney for healthcare*. Also known as a *healthcare proxy*, such an individual has the legal right to make decisions on behalf of the patient (Klein and Coogle 2015). We will further discuss this and other forms of advance directives in Chap. 9 (end-of-life ethics).

Finally, an acceptable but the least preferred level of consent derives from a patient's *best interests*. The "best interests" standard is applied in an emergency when no one is available to give legal consent. For example, consider a young man, a pedestrian struck by a hit-and-run driver, brought to the emergency room with no identification. He is unconscious and in shock from a ruptured spleen. Beneficence demands that surgeons take him to the operating room to correct the damage, on the reasonable assumption that he would give consent if able to. This idea is sometimes also called *implied consent* (Jonsen et al. 2015).

> **Levels of Consent**
> - Informed Consent by the Patient
> - Substituted Judgment (surrogate, proxy)
> - Best Interests (implied consent)

Truth Telling in Healthcare

Should a pharmacist, physician, or nurse always tell the truth, even if it seems contrary to the patient's best interests? You may recall the case of Elaina Fletcher in Chap. 2, an 82-year-old recent widow planning a tropical cruise. Her internist, Dr. Anton, considered the possibility of temporarily withholding the news of a bad test result out of compassion for her life circumstances. We pointed out that such a move would violate the Kantian doctrine of only doing an action that could be made into a rule for everyone. But other ethical theories also make this a bad idea. From a utilitarian perspective, deliberate lying or the withholding of vital information, even if well-intentioned, would inevitably lead to a breakdown in the trust relationship of healthcare professionals and their patients. Even virtue ethics demands that we avoid hedging or altering the clear, forthright delivery of accurate information. A virtuous clinician should exhibit honesty. Such a character trait is vital for personal integrity and the good of the profession (Slowther 2009).

What, if any, are the exceptions? Once again, the patient is in control, and may voluntarily choose to forego more clinical information (Jonsen et al. 2015). The following two patient statements will illustrate:

- "Doctor, I don't want to know any more details. Just do the procedure – I trust you."
- "Doctor, please just tell my husband the information – he will make all the decisions."

For pharmacists, all of this raises an interesting question: what about placebos? Is it ethical to prescribe or dispense a "sugar pill" instead of a pharmacologically-active medication? Is this ever in the patient's best interest? After all, many patients respond favorably to the "placebo effect," wherein a favorable therapeutic outcome occurs from a placebo, often with regard to a subjective symptom such as pain.[3] Such an idea almost always comes from a beneficent intent.

Here's an example: A pharmacist observes that an elderly woman is becoming lethargic as a side effect of taking 10 mg of zolpidem tartrate each night to help her sleep. When she appears at his retail pharmacy for a prescribed refill, is he justified in giving her a vitamin instead, calling it her "sleeping pill?" Aside from the legal problems with such a move, it is inherently unethical. The pharmacist should never mislead his patient but should confront the issue directly, explaining to the woman that the medication is causing a significant risk to her well-being. Misleading the patient violates the patient's ability to make autonomous decisions about her healthcare. If necessary, the pharmacist should enlist the help of the original prescriber to make the case. This approach will result in the best long-term outcome and helps to maintain a therapeutic relationship. Note that this discussion applies only to a direct clinical situation. With proper consent, placebos are often used in clinical research. We will take up the matter of research ethics in Chap. 6.

Pediatric Consent and Assent

In this final section, we turn to the unique situations that arise with minors, patients under 18 years of age. As we have seen, a minor is legally not able to give informed consent for treatment in the U.S., so parents are typically the surrogates. This standard is a natural and reasonable state of affairs since parents will *almost always* act in a child's best interests. Therefore, both ethics and law give great deference to

---

[3] The power of the placebo effect. Harvard Health Publishing Web site. www.health.harvard.edu/mental-health/the-power-of-the-placebo-effect. Published 2019. Accessed 4/9/2020.

parental wishes. This approach sets a high bar: it is always (and should be) difficult to override parents. In a select few cases, it may be necessary to ask a local court to intervene and only if parental preferences are clearly in conflict with the child's best interests. Recourse to the courts should, therefore, be rare (Katz et al. 2016).

Those who work in pediatrics recognize that children may not have the right to give consent to treatments on their own behalf. However, their *assent* is essential, especially as the child grows older. Achieving assent helps to maximize the therapeutic benefit. Obtaining assent begins with helping the child understand his or her condition at a developmentally appropriate level and telling the patient what to expect. While it may be helpful to ask the patient to express a willingness to receive treatment, this may not be appropriate if the treatment is required and will be imposed over the child's objections (Katz et al. 2016; Kuther 2003).

There are a few well-recognized exceptions to the statutory age of 18 years for consent to medical treatments. Though state laws vary, it is common to allow children 14 years of age and older to make their own decisions about sexually-transmitted infections, judgments surrounding family planning or abortion, and participation in investigational treatments.

Two other unique categories allow minors to give their consent. An *emancipated minor* lives independently of parents, due to marriage or service in the military. He or she may be treated without parental consent. Note that pregnancy does not automatically confer emancipated status, and this varies according to the state. A *mature minor* is usually age 15 or older, and appears to make reasoned judgments. In this circumstance, a proposed medical treatment is for the patient's benefit, and parental consent cannot be obtained (Jonsen et al. 2015). An interesting real-life example of this concept is that of a 14-year-old Jehovah's Witness patient with leukemia. A court allowed him to refuse a blood transfusion over his parents' objections, which eventually led to his death. The judge declared that he was a mature minor, based on his longstanding illness and his well-developed personal religious understanding (Ostrom 2007).

In summary, this chapter has begun our examination of the essential principles of clinical ethics, embedded in the historical Hippocratic healthcare tradition over two millennia. Medical principlism includes beneficence, non-maleficence, distributive justice, and the Kantian principle of autonomy. These ideas undergird their modern ethical implications for informed consent, decision-making capacity, and truthfulness. Finally, we considered how all of this plays out in the pediatric realm. Pharmacists will see applications of all of these ethical ideas as they attempt to accurately interpret prescriber medication orders, protect patients from adverse drug interactions, engage in patient education and counseling, and treat each individual with dignity and respect.

Key Terms

- beneficence
- non-maleficence
- distributive justice

- autonomy
- informed consent v. assent (in pediatrics)
- decision-making capacity v. competency
- truth-telling
- emancipated minor
- mature minor

Questions for Review and Discussion

1. Name the four components of medical principlism, and define each. Which of these ideas is the most recent?
2. What are the three possible levels of consent? Which one is the most common and preferred?
3. A 16-year-old girl comes to your retail pharmacy, requesting information on emergency contraception. Legally and ethically, do you need her parents' consent to have this discussion?
4. Why are placebos considered unethical as clinical treatments, but are considered acceptable in clinical research?
5. Are there exceptions to the truth-telling standard? On what basis should you decide?

Further Reading

Cameron, Nigel, *The New Medicine: Life and Death after Hippocrates*. Chicago, Bioethics Press, 2001. An insightful analysis of the history and sources of the original Hippocratic Oath, with applications for modern clinical practice.

Katz, Aviva and Sally A. Webb, "Informed consent in decision-making in pediatric practice." *Pediatrics 138*:2, August, 2016. A readable yet comprehensive report from the American Academy of Pediatrics, on informed consent in general, with particular emphasis on pediatric practice.

References

Autonomy. Oxford English dictionary. 2020. https://en.oxforddictionaries.com/definition/autonomy.
Beauchamp TL. Informed consent: its history, meaning, and present challenges. Camb. Q. Healthc. Ethics. 2011;20(4):515–23.
Beauchamp TL, Childress JF. Principles of biomedical ethics. 8th ed. New York: Oxford University Press; 2019.
Beneficence. Miriam-Webster Online. 2020. www.merriam-webster.com/dictionary/beneficence.
Bristow W. The enlightenment. In: Stanford encyclopedia of philosophy. 2017. https://plato.stanford.edu/entries/enlightenment/.
Bulger RJ, Barbato AL. On the Hippocratic sources of western medical practice. Hast. Cent. Rep. 2000;30(4):S4–7.
Cameron N. The new medicine: life and death after Hippocrates. Chicago: Bioethics Press; 2001.
Del Guercio LR. Triage in cold blood. Crit. Care Med. 1977;5(4):167–9.

Edelstein L. The Hippocratic oath, text, translation and interpretation. Baltimore: Johns Hopkins Press; 1943.
Edelstein L, Temkin O, Temkin CL. Ancient medicine; selected papers of Ludwig Edelstein. Baltimore: Johns Hopkins Press; 1967.
Faden RR, Beauchamp TL, King NMP. A history and theory of informed consent. New York: Oxford University Press; 1986.
Ganzini L, Volicer L, Nelson WA, Fox E, Derse AR. Ten myths about decision-making capacity. J. Am. Med. Dir. Assoc. 2005;6(3 Supplement):S100–4.
Gourevitch D. Le triangle hippocratique dans le monde gréco-romain. Le malade, sa maladie et son médecin. Vol 251: Ecoles françaises d'Athènes et de Rome; 1984.
Guinan P. Hippocrates is not dead: an anthology of Hippocratic readings. Author House; 2011.
Hippocrates. Of the epidemics. In: Internet classics archive. 400 B.C.E. / 2009. http://classics.mit.edu/Hippocrates/epidemics.html.
Hippocrates. New world encyclopedia. 2014. www.newworldencyclopedia.org/entry/Hippocrates#CITEREFNational_Library_of_Medicine2006.
Jonsen AR, Siegler M, Winslade WJ. Clinical ethics: a practical approach to ethical decisions in clinical medicine. 8th ed. New York: McGraw-Hill Education; 2015.
Katz AL, Webb SA, Bioethics Co. Informed consent in decision-making in pediatric practice. Pediatrics. 2016;2016:e20161485.
Klein ML, Coogle CL. Living wills, durable power of attorney, and advance directives. In: The encyclopedia of adulthood and aging; 2015. p. 1–5.
Kuther TL. Medical decision-making and minors: issues of consent and assent. Adolescence. 2003;38(150):343.
Ostrom C. Mount Vernon leukemia patient, 14, dies after rejecting transfusions. The Seattle Times, Nov. 29, 2007. www.seattletimes.com/seattle-news/health/mount-vernon-leukemia-patient-14-dies-after-rejecting-transfusions/.
Pope TM. Legal fundamentals of surrogate decision making. Chest. 2012;141(4):1074–81.
Requarth JA. Informed consent challenges in frail, delirious, demented, and do-not-resuscitate adult patients. J. Vasc. Interv. Radiol. 2015;26(11):1647–51.
Savel RH, Munro CL. From Asclepius to Hippocrates: the art and science of healing. In: AACN. 2014.
Slowther A. Truth-telling in health care. Clin. Ethics. 2009;4(4):173–5.
Sokol DK. "First do no harm" revisited. BMJ. 2013;347:f6426.

Chapter 5
Clinical Ethics in Historical Context, Part II

The Eugenics Era and Its Influence on Modern Healthcare Ethics

In the previous chapter, our discussion centered on the Hippocratic principles that undergird modern medical principlism, namely beneficence, non-maleficence, distributive justice, and the more recent idea of personal autonomy, which comes from the eighteenth century. As these theoretical concepts are translated into clinical practice, they retain their influence on clinical decision-making and informed consent.

From its earliest beginnings, Hippocratism has always warranted a well-founded and beneficent intent to provide medical treatments to all patients, even when those patients were slaves or other minorities (part of the definition of distributive justice). But economic and social pressures in the late nineteenth century began to erode this default commitment to equality and justice, which became very evident through a pseudoscientific movement called *eugenics*.

Eugenics derives from two Greek roots: from the Greek *eu*, meaning "good," and *genos*, meaning "birth."[1] The underlying ideas are quite ancient in origin. For example, Plato argued in his *Republic* that human baby production should be limited to people selected for desirable qualities, with marriage and reproduction controlled by the state.[2] The word eugenics was coined in 1883 by Francis Galton, an Englishman and cousin of Charles Darwin. He wanted to apply Darwinian science to ideas of heredity and the improvement of humanity. Daniel Kevles has said this: "Since Galton's day, 'eugenics' has become a word of ugly connotations – and deservedly" (Kevles 1985).

---

[1] Eugenics. Online Etymology Dictionary. www.etymonline.com/word/eugenics. Published 2018.
[2] Goering S. Eugenics. Stanford Encyclopedia of Philosophy. https://plato.stanford.edu/entries/eugenics/. Published 2014.

In the beginning, Galton had the best of intentions. A British scientist, he argued that genius and talent are inherited and that we could improve our future generations by encouraging the "best" in society to have more children. Such "positive eugenics" seems innocent enough. After all, most of us want to "marry well." Which one of us would like to find a spouse whom we don't look up to and admire? Likewise, most of us want to have intelligent and attractive children (in fact, we may bristle at the idea that they are not!). But that was not the outcome of this movement in the early twentieth century, which soon degenerated into "negative eugenics," mandating forced sterilization and even the elimination of undesirables. Kevles again: "By 1935 eugenics had become hopelessly perverted into a pseudoscientific facade for advocates of race and class prejudice, defenders of vested interests of church and state, Fascists, Hitlerites, and reactionaries generally" (Kevles 1985).

The societal context of all of eugenics is vital for us to understand. The Industrial Revolution had caused a rapid growth of industry, with its increased mechanization of agriculture and significant population shifts. Since rural workers were leaving the farms for cities in droves, cities expanded much faster than available housing, leading to overcrowding and the first urban slums. Labor unions arose due to the widespread exploitation of labor, and there was a massive influx of immigrants from Europe in the years before World War I.[3]

Against this economic and social backdrop the Darwinian Revolution emerged, which began in 1859 with the publication of Charles Darwin's seminal work, *On the Origin of Species* (Darwin 1859). Along with its impact on science, Darwin's theory fostered a more dubious influence on social norms, that of *Social Darwinism*. Social Darwinism attempted to explain social and economic inequalities as a necessary result of the "survival of the fittest." The birth rate among the wealthy was declining, while the working-class reproduced at a much faster rate. This situation created alarm within the upper class, along with concerns about overcrowding and limited resources. Social philanthropy and religious institutions were little help; instead, progressive reformers looked to science as a "cure-all" for the ills of society, and eugenics societies sprang up throughout Great Britain and the United States.[4]

This new form of social engineering taught that there were certain "genetically selected" traits, such as pauperism, feeble-mindedness, alcoholism, rebelliousness, nomadism, criminality, and prostitution, all due to "defective germ plasm." If all of this seems vague, even ridiculous, by today's standards, remember that modern ideas of genetics were unknown at the turn of the twentieth century. Gregor Mendel, the Austrian monk and botanist, had performed his well-known pea-breeding experiments between 1854 and 1863, presenting his conclusions before a scientific body in 1865. However, his research was unknown to Charles Darwin and was mostly forgotten until well into the new century.[5]

---

[3] Industrial Revolution. History.com. www.history.com/topics/industrial-revolution. Published 2018.

[4] Image Archive on the American Eugenics Movement. Cold Spring Harbor Laboratory Web site. www.eugenicsarchive.org/eugenics/. Published 2018. Accessed 2/15/2021.

[5] Gregor Mendel. Biography Web site. www.biography.com/people/gregor-mendel-39282. Published 2018. Accessed 2/15/2021.

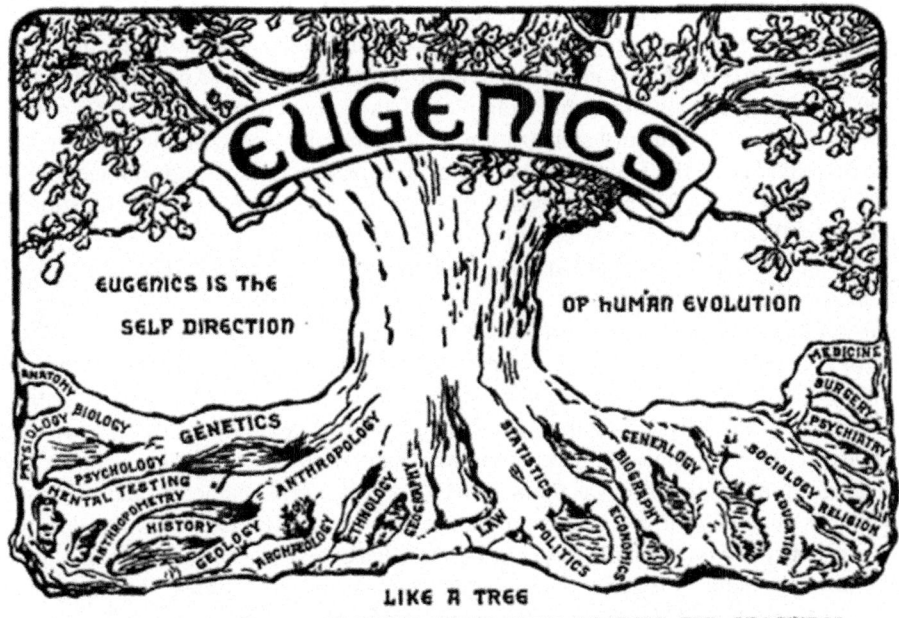

Fig. 5.1 Popular image for eugenics, showing its multiple influences. (Image Public Domain, https://commons.wikimedia.org/w/index.php?curid=135048)

Against this background, eugenicist Charles Davenport established the Eugenics Record Office (ERO) in 1910, located in Cold Harbor, New York. It notably included Alexander Graham Bell as chairman of its Board of Scientific Directors.[6,7] The ERO trained field workers to collect pedigrees of families with interesting traits, including pauperism, criminality, and feeble-mindedness. For example, they described "naval officer" as an inherited trait, composed of sub-traits for "thalassophilia" (love of the sea) and "wanderlust," concluding that the trait is unique to males. These ideas may appear rather quaint and outdated, yet they had substantial legal weight. One of the purported goals of the ERO was "to advise concerning the eugenical fitness of proposed marriages" (Fig. 5.1).[8]

As eugenics became popular in the early twentieth century, it led to aggressive attempts to promote it. In the 1920s, the American Eugenics Society sponsored

---

[6] Image Archive on the American Eugenics Movement. Cold Spring Harbor Laboratory Web site. www.eugenicsarchive.org/eugenics/. Published 2018. Accessed 2/15/2021.

[7] Krisch JA. When Racism Was a Science. New York Times. www.nytimes.com/2014/10/14/science/haunted-files-the-eugenics-record-office-recreates-a-dark-time-in-a-laboratorys-past.html. Published Oct. 13, 2014.

[8] Image Archive on the American Eugenics Movement. Cold Spring Harbor Laboratory Web site. www.eugenicsarchive.org/eugenics/. Published 2018. Accessed 2/15/2021.

"fitter family" and "better baby" contests, with prizes given to the winners. Such examples of "positive eugenics" may seem benign, but they had a darker side, promoting racism by marginalizing foreign immigrants and African Americans (Pernick 2002). But there were also many examples of "negative eugenics." A 1907 Indiana law mandating compulsory sterilization of "degenerates" became the first eugenic sterilization law in the United States. Other sterilization laws would follow in other states. At one point, 33 states had passed laws permitting involuntary sterilization of mentally ill individuals or those with cognitive disabilities.[9]

All of this came to a startling culmination with the case of *Buck v. Bell*, 274 U.S. 200 (1927), decided by the U.S. Supreme Court in 1927. A brief overview of this landmark decision will show how its repercussions are still with us today.[10]

In 1924, Carrie Buck was a 17-year-old child who grew up with foster parents in Charlottesville, Virginia. Her mother, Emma Buck, had long before been institutionalized in the Virginia Colony for Epileptics and Feeble-Minded in Lynchburg (though colony officials classified Emma as feebleminded, the real reason for her incarceration was that she was thought to be "of low moral character" and on public welfare). When Carrie became pregnant out of wedlock, her foster parents decided to put her away and sought the help of the Charlottesville Department of Public Welfare. A judge and two doctors determined that Carrie was feebleminded (Fig. 5.2) (Cohen 2016).

In the 1920s, feeblemindedness was a vague diagnosis, seemingly related to a mental or social defect. Feebleminded people were thought to be a menace to the social order and were "a threat to the genetic health and stability of the race" (Kline 2001). Amazingly, administrators cited no specific test to document a cognitive disability on Carrie's part, and her grades in school were adequate. The doctors, contrary to any evidence whatsoever, also declared her to be epileptic. In their petition, Carrie's parents hid the fact that she was pregnant. It would later be revealed that a family member had raped her, clearly the most likely reason that her foster parents wanted her quietly put away. After sending her to the Virginia Colony, Carrie's baby Vivian was born, who was taken care of by Carrie's foster parents (Cohen 2016).[11]

In March of that same year (1924), the Virginia legislature had passed the "Eugenical Sterilization Act," which allowed state institutions to involuntarily sterilize particular residents "afflicted with hereditary forms of insanity that are recurrent, idiocy, imbecility, feeble-mindedness, or epilepsy" (Landman 1932). Officials of the state hospital system were looking for a test case to validate the law as constitutional before carrying out any of its provisions (Cohen 2016). They found their candidate in Carrie Buck, whom they were determined to sterilize.

---

[9] Image Archive on the American Eugenics Movement. Cold Spring Harbor Laboratory Web site. www.eugenicsarchive.org/eugenics/. Published 2018. Accessed 2/15/2021.

[10] Many of the details in this narrative come from: Adam Cohen, *Imbeciles: the Supreme Court, American Eugenics, and the Sterilization of Carrie Buck* (2016). New York, Penguin Press.

[11] Image Archive on the American Eugenics Movement. Cold Spring Harbor Laboratory Web site. www.eugenicsarchive.org/eugenics/. Published 2018. Accessed 2/15/2021.

The Eugenics Era and Its Influence on Modern Healthcare Ethics

Fig. 5.2 Carrie and Emma Buck at the Virginia Colony for Epileptics and Feebleminded. (Image courtesy of Arthur Estabrook Papers, M.E. Grenander Special Collections & Archives, University at Albany, SUNY)

As they went to the courts, a field worker from the ERO testified in the first trial as an expert witness, based on his examination of Emma (the mother), her daughter Carrie, and Carrie's baby Vivian. In his estimation, all were feebleminded or likely to become so, and such feeblemindedness was an inherited trait.[12] Because of alleged antisocial conduct by Carrie and her mother, this would most likely lead to criminality and pauperism, and Carrie would continue to be a burden on society. A weak defense by Carrie's appointed lawyer claimed protection under the 14th Amendment, which specified that no state should "deprive any person of life, liberty, or property, without due process of law."[13] However, the state's lawyers compared compulsory sterilization to compulsory vaccination for the promotion of public health, which won over the lower courts, setting up the final hearing in the U.S. Supreme Court.

The High Court rendered its verdict in *Buck v. Bell* on May 2, 1927. By an 8-1 margin, the justices endorsed the constitutionality of Virginia's sterilization law, with the majority opinion rendered by Oliver Wendell Holmes, Jr. His ruling stated, in part:

---

[12] Buck v. Bell: The Test Case for Virginia's Eugenical Sterilization Act. University of Virginia Historical Collections Web site. http://exhibits.hsl.virginia.edu/eugenics/3-buckvbell/. Published 2007. Accessed 2/15/2021.

[13] Fourteenth Amendment. Encyclopædia Britannica. www.britannica.com/topic/Fourteenth-Amendment. Published 2018.

It is better for all the world if, instead of waiting to execute degenerate offspring for crime or to let them starve for their imbecility, society can prevent those who are manifestly unfit from continuing their kind. The principle that sustains compulsory vaccination is broad enough to cover cutting the Fallopian tubes ... Three generations of imbeciles are enough.[14]

Virginia Colony Superintendent John Bell sterilized Carrie Buck on October 19, 1927.

Harry Laughlin, who had been the author of the "model sterilization act" in Virginia, made a draft available to state and foreign governments. The statute subsequently served as a model for Germany's Hereditary Health Law of 1933. At Nuremberg, Nazi lawyers cited the *Buck v. Bell* decision as an acceptable precedent for the sterilization of 2 million people in its "racial hygiene" program. Virginia's law also served as a model for similar legislation in 30 other U.S. states. In all, over 60,000 U.S. citizens have undergone involuntary sterilization based on eugenics laws.

By the mid-1930s, widespread abuses by the Nazis caused many to lose their enthusiasm for eugenics. The emergence of Mendelian genetics led to a more scientific understanding of heredity and inheritance, which undermined much of the work of the ERO. At the same time, political support for the movement also declined. In 1935, a scientific committee found the work of the ERO to be "without scientific merit," and the office quietly closed its doors on December 31, 1939.

The *Buck v. Bell* decision is considered one of the worst legal precedents in history; the U.S. Supreme Court has never officially reversed it (Kevles 1985; Cohen 2016).[15]

The Era of Nazi Medicine

As far back as the 1890s, eugenics ideas were very influential in Germany, even as social Darwinism became a movement in the U.S. and England. In 1896, Alfred Ploetz had introduced the term *rassenhygiene* (racial hygiene), which greatly influenced German ideas of national purity (Weindling 1985). During World War I, for example, rationing caused many psychiatric patients in German hospitals to die of starvation due to their low priority (Torrey and Yolken 2010). These pressures led to a more rapid reversion to the negative aspects of eugenics than it did in other countries (Fig. 5.3).

The most influential philosophical justification for negative eugenics in Germany was the 1920 publication of *Permission to Destroy Life Unworthy of Life*, by Karl Binding and Alfred Hoche (Baker and McCullough 2007). They argued that *lebensunwertes leben*, German for "life unworthy of life," justified medical killing. For

---

[14] Holmes OW. *Buck v. Bell*, 274 U.S. 200 (1927). In: Justia: U.S. Supreme Court; 2018.
[15] Image Archive on the American Eugenics Movement. Cold Spring Harbor Laboratory Web site. www.eugenicsarchive.org/eugenics/. Published 2018. Accessed 2/15/2021.

Fig. 5.3 Dr. Karl Brandt as a defendant at Nuremberg. (Image is Public Domain. Portrait of Karl Brandt as a defendant in the Medical Case Trial at Nuremberg. United States Holocaust Memorial Museum, Photograph #06231)

them, the right to life must be *earned*, not assumed. This idea is a chilling extension of social Darwinist thinking and "survival of the fittest" (Meyer 1988). These concepts became part of the national psyche as Germany began to recover its national pride on the heels of its humiliating defeat in World War I. Therefore, as Schaefer put it, "public opposition to eugenics was virtually non-existent, and as the Nazis began to take over in the 1930s, there was little opportunity for dissent" (Schaefer 2004). The ensuing years saw a widespread recognition of some individuals as *minderwertig* (inferior). German society widely debated the ideas of Binding and Hoche, with an increased acceptance of eliminating the disabled as "useless eaters" (Mostert 2002).

During this time, eugenics considerations and an emphasis on German racial purity led to the sterilization of 2 million German citizens under the Hereditary Health Law of 1933, as mentioned earlier. Coercive sterilization gave way to extermination: euthanasia of "impaired" children in hospitals (1939), euthanasia of "impaired" adults in mental hospitals with carbon monoxide gas (1940), and euthanasia of "impaired" inmates in concentration camps (1941) (Meyer 1988; Schaefer 2004; Caplan 2012; Lifton 2000).

The German state's broader political agenda and its bias against an entire race had already become evident by 1935. Laws enacted in Nuremberg excluded Jews from citizenship and expressly outlawed sexual intimacy between Jews and Germans. It should, therefore, be no surprise that even political "undesirables" such as Jews, Poles, and Gypsies would later be seen as *minderwertig*, leading eventually to Goering's genocidal Final Solution.[16]

Euthanasia measures became common without the knowledge of the German people, but many began to suspect what was going on. As a result, the German

---

[16] History.com. Goering orders Heydrich to prepare for the Final Solution. www.history.com/this-day-in-history/goering-orders-heydrich-to-prepare-for-the-final-solution. Published 2018.

propaganda machine tried certain tactics to appease the populace. One example was the film *Ich klage an* (English: *I Accuse*), in which a physician gives a lethal injection to his incurably ill wife, who begs him to do it. During his trial for murder, the defense position was that the patient asked for it. Seen by over 15 million Germans, the film was even awarded a prize at the Venice Biennale Festival and created a substantial public discussion and sympathy for the legalization of euthanasia (mercy killing) (Michalczyk 1994). Ironically, the sympathetic portrayal of euthanasia depicted in the film *Ich klage an* had no resemblance to actual German practices, which were involuntary and against the wishes of both victims and families. Despite such attempts at propaganda, a growing public outcry stopped the practice of gassing mental patients in hospitals, and officials transferred equipment for this purpose to the death camps.

It is far beyond the purpose of this chapter to further document the Holocaust of World War II between 1941 and 1945, which specifically targeted six million Jews for extermination, but also led to the deaths of millions of other ethnic and political groups (Landau 2016). For our purposes, we will restrict our discussion to the medical ethics arena. We will specifically consider how Nazi medicine and its perverse experiments on innocent human beings arose directly from eugenics ideas. By way of prologue, consider the late Elie Wiesel, Holocaust survivor and eloquent humanist, who asks some profoundly important questions:

> Inspired by Nazi ideology and implemented by its apostles, eugenics and euthanasia in the late 1930s and early 1940s served no social necessity and had no scientific justification. Like a poison, they ultimately contaminated all intellectual activity in Germany. But the doctors were the precursors. How can we explain their betrayal? What made them forget or eclipse the Hippocratic Oath? What gagged their conscience? What happened to their humanity? (Wiesel 2005)

We will offer our response to Dr. Wiesel and his plaintive questions at the end of this chapter. For now, we will summarize how Nazi ideology influenced medical practice and research on prisoners in the death camps. We wish to pose three questions concerning this dark period in medical history: (1) Why did Nazi physicians perpetrate such evil in the name of scientific progress? (2) What ethical lessons can we learn? Finally, (3) Can we justify utilizing any of the unethically collected data, or should we completely discard it?

We begin our analysis with the overall goals of medical experiments in the death camps. Physician David Bogod provides a helpful overview:

> The Nazi ideology was predicated on the concept of racial supremacy. At the top of the tree was the Aryan race; at the foot were the 'untermenschen:' blacks, gypsies, homosexuals, and Jews. In the obscene logic which emerged from this categorisation, such 'subhumans' were legitimate targets for extermination and, before their deaths, experimentation. A small number of German doctors who espoused the Nazi philosophy – some, like the infamous Joseph Mengele, highly regarded and well-published – were recruited to carry out medical experiments on concentration camp inmates (Bogod 2004).

Based on this agenda, the Nazi doctors performed experiments they thought would be relevant to the war effort. For example, one line of investigation proposed to find the best method to improve the survival of Luftwaffe pilots forced to bail out over the frigid waters of the North Sea. For this purpose, Dr. Sigmund Rascher, between August of 1942 and May 1943, immersed 300 prisoners in ice water, either against their will or with false promises of being released from prison. The experiments tested different methods of rewarming the subjects, though most of them died from the induced hypothermia (Bogod 2004; Berger 1990).

Another example of war-related experiments also relates to German pilots, who often had to fly up to high elevations to avoid Allied aircraft. When forced to bail out at such high altitudes, the pilots were subjected to very low atmospheric pressures, which could result in decompression sickness and death. To study this phenomenon, Dr. Rasher tested about 200 prisoners by placing them into chambers that simulated heights of 20,000–40,000 feet above sea level, with associated low oxygen levels and low pressures. Most of the subjects died from the experiments, and doctors performed autopsies to determine the effects on brains, lungs, and hearts (Mitscherlich and Mielke 2015).

Extensive documentation provides evidence of many other cases of abuse, all with the rationale of providing scientific information to help the war effort. The doctors forced some prisoners to drink seawater to assess various methods to treat dehydration. Others were injected with malaria parasites, typhus or tuberculosis bacteria, or were given viral hepatitis. Still other victims suffered from deliberate surgical wounds without anesthesia, which were then contaminated to test the efficacy of the drug sulfanilamide. Even more shocking were reports that Himmler was seeking an anticoagulant to treat battlefield wounds. He requested experiments to perform amputations on conscious subjects, testing various agents by counting and timing drops of blood from the freshly-cut stumps (Baker and McCullough 2007; Schaefer 2004; Caplan 2012; Lifton 2000; Mitscherlich and Mielke 2015; Spitz 2005).

In some cases, there was not even a minimal attempt at any justification for the brutal practices. Medical students and young doctors who wanted to gain surgical experience in their specialty were permitted to choose healthy subjects among the prisoners and perform mutilating procedures. Unnecessary hysterectomies, Caesarian sections, and amputations were common (Lifton 2000).

As reported by Robert Lifton, a physician survivor of the Holocaust once asked a Nazi doctor about his participation in horrible medical experiments and the extermination of Jews: "How can you reconcile that with your [Hippocratic] oath as a doctor?" The doctor's reply was chilling:

> Of course I am a doctor and I want to preserve life. And out of respect for human life, I would remove a gangrenous appendix from a diseased body. The Jew is the gangrenous appendix in the body of mankind (Lifton 2000).

And so we can see the final result of substituting a physician's ethical duty to individual patients with a commitment to the aims and goals of the State. Fueled by the flawed ideology of social Darwinism and eugenics, the Third Reich thought to transform the state into a perfect Aryan ideal at the expense of individuals, conveniently scapegoating Jews and other undesirables in the process. This agenda led the Nazis to regard the marginal groups as nothing more than discardable vermin.

After the end of World War II, the Nuremberg Trials were conducted as a series of military tribunals to prosecute German war criminals, led by judges from four countries. As a part of these overall proceedings, the Doctors' Trial (also known as the Medical Case) began on December 9, 1946. This first tribunal, presided over by American judges, brought criminal proceedings against 23 German physicians and administrators for war crimes and crimes against humanity. The process lasted for 140 days, including testimony from 85 witnesses and the submission of 1500 documents. The American judges reached their verdict on August 20, 1947, pronouncing sixteen of the doctors guilty, and many received lengthy prison terms. Seven were sentenced to death and were executed by hanging on June 2, 1948 (Lifton 2000; Spitz 2005; Annas and Grodin 2018).

What ethical conclusions can we derive from this terrible chapter in history? Let us return to the questions we posed earlier and try to respond to Dr. Elie Wiesel. First, why did Nazi physicians perpetrate such evil in the name of scientific progress? In our opinion, they had lost their scientific objectivity. In viewing *lebensunwertes leben* as "medical therapy," they made the State and its goals more important than individual human lives. They subjugated their default, normative medical humanity to the glory of the Third Reich. Their thinking became bureaucratic and almost unconsciously automatic. Following the war, philosopher Hannah Arendt would refer to this idea as the "banality of evil" (Arendt 1963). Something *banal* is "insipid" or "unoriginal."[17] This form of evil is almost more horrific than the evil of hatred; many of the Nazis simply regarded the Jewish people as *unworthy of their notice*, so complete was their corruption of personal value for an entire race.

In this context, it is worth remembering the two frameworks for human value that we examined in Chap. 3, *ontological personalism* and *empirical functionalism*. Ontological personalism assumes that all human beings have intrinsic value. By contrast, empirical functionalism determines human value by one's functions or abilities. The Third Reich took the functionalist view to an extreme: the only value of a Jewish or Polish prisoner was the *work* that he or she could do. The entry gate into many of the Nazi death camps, such as Auschwitz, Sobibor, and Treblinka, had this sign emblazoned over it: *"Arbeit Macht Frei,"* which means "Work Makes You Free." So the prisoners were simply worked to death; that was their only purpose in the view of the State (Roth 1980).

Second, what ethical lessons can we learn? There are many, but we will focus primarily on one. Aside from the obvious, innumerable examples of torture and abuses of autonomy, we must never forget how the Nazi doctors perverted their Oath. They tried to make it apply to the nation as a whole, to the vision of *rassenhygiene* that drove them to an Aryan vision of a perfect State.

As we discussed in Chap. 4, the Hippocratic tradition and medical principlism protect individual patients. Beneficence, non-maleficence, distributive justice, and autonomy are all *personal* terms, and should only refer to separate and specific

---

[17] Banal. Meriam-Webster Online Dictionary. www.merriam-webster.com/dictionary/banal. Published 2018.

human beings. Employing these values in a broader context is always highly problematic. Clinicians must first and foremost address the medical and ethical issues of the individual patient in front of them, not some higher agenda.

Finally, can we justify utilizing any of the unethically collected data, or should it be discarded? There are many sides to this complex question, and we can only summarize some of the main arguments here. On the one hand, some might argue from a purely utilitarian perspective that any useful data, even if unethically obtained, might help humanity in the future. Consider, for example, that the cold-water immersion experiments at Dachau could be used to help prevent immersion-related deaths and near-drowning in modern patients (Bogod 2004). Others might employ a Kantian framework to argue that such use of the Nazi data merely perpetuates a flawed maxim, and that it could never pass Kant's Categorical Imperative test (refer to Chap. 2 to review these ethical theories).

Note that one could attempt to sidestep the ethical dilemma by merely positing that the Nazi experiments were vague and poorly controlled, and therefore lacked scientific rigor. Nonetheless, some researchers have attempted to access it and use it for a variety of modern scientific and medical purposes. One suggestion would be to grant permission to use all of the data and let the scientific community evaluate its worth. Another approach suggests asking the victims and their descendants what they wish (Baker and McCullough 2007; Schaefer 2004; Caplan 2012; Lifton 2000; Berger 1990; Mitscherlich and Mielke 2015; Spitz 2005). This entire debate is complex and remains unresolved today; one's conclusions may depend on the ethical frame or lens used in the argument.

In summary, this chapter has examined two significant areas of clinical ethics, primarily from the twentieth century. We began with a discussion of the flawed pseudo-science of eugenics and showed how it eventually was found both ethically suspect and scientifically without merit. Nonetheless, the social Darwinist philosophies undergirding eugenics reached their full development in Nazi Germany and helped play a role in the Holocaust itself.

We will cover much more detail on the ethics of human experimentation and how this differs from clinical therapies. To accomplish this task, we will undertake a parallel description of other major human research abuses in the twentieth century and how they were eventually addressed and curtailed. These matters will be the subject of the next chapter.

Key Terms

- eugenics: positive and negative
- Francis Galton
- social Darwinism
- Eugenics Record Office (ERO)
- *Buck v. Bell* Supreme Court case of 1927
- *rassenhygiene*
- *lebensunwertes leben*
- euthanasia
- Nuremberg Trials

Questions for Review and Discussion

1. Define social Darwinism. Explain both positive and negative eugenics. What social and economic factors in the U.S. led to the rise of eugenics?
2. Describe the *Buck v. Bell* Supreme Court case of 1927. What was the outcome? What far-reaching effects did this case have on eugenics practice and laws, both in the U.S. and abroad?
3. How was the Nazi practice of "euthanasia" different from euthanasia as it might be practiced today?
4. Give examples of how the Nazi doctors violated all four rules of medical principlism.
5. Should we utilize Nazi research data for some modern-day medical or research applications? Give one argument on each side of the question.

Further Reading

Arthur Caplan (Ed.), *When Medicine Went Mad: Bioethics and the Holocaust* (Caplan 1992). Totowa, New Jersey, Humana Press, 2012. Essential reading for a deeper historical understanding of the misguided and horrible medical experiments performed on innocent victims by the Third Reich. Especially moving are the personal stories by Holocaust survivors and family members.

Cold Spring Harbor Laboratory: *Image Archive on the American Eugenics Movement* (Web resource). A content-rich source of over 2500 images, including articles, photographs, charts, brochures, pedigrees, news articles, and other historical artifacts, all documenting the history of the eugenics movement from its earliest beginnings to the present. Available at: www.eugenicsarchive.org/eugenics.

Robert Lifton, *The Nazi Doctors: Medical Killing and the Psychology of Genocide*. New York, Basic Books, 2000 (Lifton 2000). The masterwork of a psychiatrist who interviewed doctors, administrators, victims, and families to provide a thoroughly convincing analysis of the motivations of the Nazi healthcare professionals who lost their way. A disturbing yet surprisingly hopeful read.

References

Annas GJ, Grodin MA. Reflections on the 70th anniversary of the Nuremberg doctors' trial. Am. J. Public Health. 2018;108(1):10–2.

Arendt H. Eichmann in Jerusalem. New York: Viking Press; 1963.

Baker R, McCullough LB. Medical ethics' appropriation of moral philosophy: the case of the sympathetic and the unsympathetic physician. Kennedy Inst. Ethics J. 2007;17(1):3–22.

Berger RL. Nazi science – the Dachau hypothermia experiments. N. Engl. J. Med. 1990;322(20):1435–40.

Bogod D. The Nazi hypothermia experiments: forbidden data? Anaesthesia. 2004;59(12):1155–6.

Caplan AL. When medicine went mad: bioethics and the Holocaust. Totowa: Humana Press; 1992.

Caplan A. When medicine went mad: bioethics and the Holocaust. Totowa: Human Press; 2012.

Cohen A. Imbeciles: the Supreme Court, American eugenics, and the sterilization of Carrie Buck. New York: Penguin Press; 2016.

References

Darwin C. On the origin of species by means of natural selection. London: J. Murray; 1859.

Kevles DJ. In the name of eugenics: genetics and the uses of human heredity. 1st ed. New York: Knopf; 1985.

Kline W. Building a better race: gender, sexuality, and eugenics from the turn of the century to the baby boom. Berkeley: University of California Press; 2001.

Landau RS. The Nazi Holocaust: its history and meaning. New York: IB Tauris; 2016.

Landman JH. Human sterilization; the history of the sexual sterilization movement. New York: Macmillan; 1932.

Lifton RJ. The Nazi doctors: medical killing and the psychology of genocide. New York: Basic Books; 2000.

Meyer J-E. The fate of the mentally ill in Germany during the Third Reich. Psychol. Med. 1988;18(3):575–81.

Michalczyk JJ. Euthanasia in Nazi propaganda films: selling murder. In: Medicine, ethics, and the Third Reich: historical and contemporary issues. 1994. pp. 64–70.

Mitscherlich A, Mielke F. Doctors of infamy: the story of the Nazi medical crimes. Pickle Partners Publishing; 2015.

Mostert MP. Useless eaters: disability as genocidal marker in Nazi Germany. J. Spec. Educ. 2002;36(3):155.

Pernick MS. Taking better baby contests seriously. Am. J. Public Health. 2002;92(5):707–8.

Roth JK. Holocaust business: some reflections on arbeit macht frei. Ann. Am. Acad. Pol. Soc. Sci. 1980;450(1):68–82.

Schaefer N. The legacy of Nazi medicine. The New Atlantis. 2004;5:54–60.

Spitz V. Doctors from hell: the horrific account of Nazi experiments on humans. 1st Sentient Publications ed. Boulder: Sentient Publications; 2005.

Torrey EF, Yolken RH. Psychiatric genocide: Nazi attempts to eradicate schizophrenia. Schizophr. Bull. 2010;36(1):26–32.

Weindling P. Weimar Eugenics: the Kaiser Wilhelm Institute for Anthropology, human heredity and eugenics in social context. Ann. Sci. 1985;42(3):303.

Wiesel E. Without conscience. N. Engl. J. Med. 2005;352(15):1511–3.

Chapter 6
Clinical Ethics in Historical Context, Part III

Human Research Ethics: The Tuskegee Syphilis Study

As we saw in the preceding chapter, the highly questionable theory of eugenics began to fall out of favor in the United States in the 1930s, for two main reasons: (1) the increasing scientific sophistication of genetics made previous eugenics assumptions untenable, and (2) the terrible abuses of the Nazis cast a moral cloud over eugenic philosophy. However, the movement left behind many subtle social attitudes and biases. It helped to strengthen a strong undercurrent of racism and classism that had long existed in the U.S. since the Civil War and had persisted in the aftermath of World War I. These attitudes would not be completely and publically revealed until the middle of the twentieth century and the civil rights movement.

Against this backdrop, the Great Depression opened the door for a dark chapter in the history of U.S. healthcare and medicine. The abuses at Tuskegee were an additional major departure from Hippocratic principles and eventually led to significant reforms. Consider the following summary statement:

> In late July of 1972, Jean Heller of the Associated Press broke the story: for forty years, the United States Public Health Service had been conducting a study of the effects of untreated syphilis on black men in Macon County, Alabama, in and around the county seat of Tuskegee. The Tuskegee Study, as the experiment had come to be called, involved a substantial number of men: 399 who had syphilis and an additional 201 who were free of the disease chosen to serve as controls. All of the syphilitic men were in the late stage of the disease when the study began (Jones 1981).

Thus begins *Bad Blood: The Tuskegee Syphilis Experiment*, a landmark exposé by James H. Jones, a research fellow at the Kennedy Institute of Ethics at Georgetown University. This meticulously-documented cautionary text, as well as an earlier analysis by Allan Brandt of the Hastings Center, are the primary sources for the discussion in this chapter (Jones 1981; Brandt 1978).

In 1929, the Julius Rosenwald Fund provided a grant that allowed the United States Public Health Service (USPHS) to study six counties in rural Alabama to determine the prevalence of syphilis among blacks and the possibilities for widespread treatment. Tuskegee is located in Macon County, Alabama, and had the highest incidence of the disease among the counties studied. The USPHS had already concluded that measures to treat syphilis should be instituted there, though the massive economic downturn meant that these plans were never implemented (Brandt 1978).

Three years later, in 1932, Dr. Taliaferro Clark, the head of the USPHS Venereal Disease Division, concluded that conditions in Macon County warranted further study and proposed recruiting black men with advanced syphilis to undergo a series of periodic examinations. His sole purpose was to study the "natural history" of the disease, even though some treatments were available. Existing treatments at the time (certain arsenical compounds and bismuth injections), though toxic, were still recommended as better than nothing (Brandt 1978). Dr. Clark reportedly wrote to one of his colleagues:

> Macon County is a natural laboratory; a ready-made situation. The rather low intelligence of the Negro population, depressed economic conditions, and the common promiscuous sex relations not only contribute to the spread of syphilis but the prevailing indifference with regard to treatment.[1]

Once known as the "great imitator," syphilis is a sexually-transmitted disease that can present in complex and confusing ways. The *primary lesion*, called a chancre, is a painless genital ulcer that appears about 3 weeks after exposure. Untreated cases give way to *secondary syphilis*, characterized by fever, a generalized rash, and enlarged lymph nodes. Years later, the most severe outcomes of *tertiary syphilis* are aortic and vascular disease, as well as neurological complications. Severe neurosyphilis may lead to ataxia, seizures, dementia, and paralysis (Brown and Frank 2003).

With syphilis poorly understood back in 1932, Dr. Clark and the USPHS mounted a campaign to find black men in Macon County with the disease. They posted handbills offering "free blood tests" and "free treatment" to "colored people with bad blood." Eventually, they enrolled 399 men with a positive Wasserman test for syphilis and 201 men without the disease (Fig. 6.1) (Brandt 1978).

Thus began a 40-year program of deception and misinformation. The inherent racism of the physicians and researchers led them to believe that the study subjects were ignorant and incapable of truly understanding their diagnosis. They never told them the true nature of their illness, nor were they ever offered any meaningful treatment; it was all about the clinical observations of untreated disease. For example, they performed painful lumbar punctures on the subjects to establish a diagnosis of neurosyphilis. Though strictly diagnostic, researchers misled the men into thinking that some form of therapy was involved. Here is a letter sent to the participants:

---

[1] Sundararajan N, Kao A, Sood A. Bad Blood: The Tuskegee Syphilis Study. http://tuskegeestudy.weebly.com/macon-county.html. Published 2014. Accessed 2/15/2021.

Human Research Ethics: The Tuskegee Syphilis Study 65

Fig. 6.1 A blood draw in a test subject at Tuskegee. (Public Domain: National Archives, https://catalog.archives.gov/id/956117)

> Some time ago you were given a thorough examination and since that time we hope you have gotten a great deal of treatment for bad blood. You will now be given your last chance to get a second examination. This examination is a very special one and after it is finished you will be given a special treatment if it is believed you are in a condition to stand it... REMEMBER THIS IS YOUR LAST CHANCE FOR SPECIAL FREE TREATMENT. BE SURE TO MEET THE NURSE (emphasis in the original) (Brandt 1978).

The ultimate goal of the program was to get all deceased participants into an autopsy, which the Brandt report makes very clear. To ensure that severely ill subjects went to the hospital, the USPHS offered to pay their funeral expenses (Brandt 1978).

The purpose of the study to observe and not treat the research subjects was reinforced during World War II. In 1941, The U.S. Army attempted to draft several of them, but the USPHS submitted names of 256 Tuskegee participants to have them excluded. If the military had succeeded in inducting these men into the Army, this would have led to physical exams and eventual treatment, which would have interfered with the aims of the project. Ironically, penicillin became widely available during the 1940s and might have led to some therapeutic benefit if utilized. Ultimately, the Tuskegee syphilis study lasted over 40 years, and its compiled data formed the basis for 13 research reports in peer-reviewed medical journals (Brandt 1978).

All of this began to unravel in the late 1960s. Social worker Peter Buxtun went to work as a venereal disease investigator for the Public Health Service in San Francisco in 1965, and soon after that, he became aware of the Tuskegee project. He was appalled, and in 1966 wrote a letter to the director of the Division of Venereal Diseases of the CDC (Center for Disease Control). The CDC ignored his message, so Buxtun persisted in his complaint. A subsequent meeting with CDC officials in

San Francisco was merely an attempt to intimidate him, and nothing changed, so in 1967 Buxtun resigned from the Public Health Service. A second letter by Buxtun in 1968 finally led to results, and the CDC held a series of meetings to discuss the Tuskegee study. In the end, it became apparent that none of the key officials were willing to stop the project based on ethical concerns because they felt there was ongoing scientific value to it (Jones 1981).

Finally, in anger and frustration, Peter Buxtun went to the Associated Press (A.P.) in San Francisco in the summer of 1972, where his story received a much more sympathetic hearing. A.P. reporter Jean Heller did additional research, then broke her bombshell story in the *Washington Star* on July 25, 1972 (Jones 1981). A widespread public outcry led to the official closure of the study by November of that year.[2] In July of the following year, civil rights attorney Fred Gray filed a class-action lawsuit on behalf of the study subjects and their descendants, a case that never made it to trial. In December of 1974, the U.S. government agreed to an out-of-court settlement of $10 million (Jones 1981).

Finally, on May 16, 1997, President Bill Clinton met with several of the remaining survivors of the Tuskegee study in the East Room of the White House, along with family members and descendants. His official apology included these words:

> The people who ran the study at Tuskegee diminished the stature of man by abandoning the most basic ethical precepts. They forgot their pledge to heal and repair. They had the power to heal the survivors and all the others and they did not. Today, all we can do is apologize.[3]

What is the legacy of Tuskegee? At the very least, Tuskegee is a cautionary case study in medical ethics. The researchers, aided and abetted by the USPHS, violated all of the precepts of medical principlism we discussed in Chap. 4: *beneficence*, because there was no direct benefit to the African-American participants and their families, and *non-maleficence*, because the withholding of treatment led to direct harms to the men, their families, and their descendants. The study violated *distributive justice* because racist assumptions undergirded the project from the beginning. Racism and classism persisted in the project's disrespectful treatment of poor and uneducated rural men. And of course, there was no respect for *autonomy*. By not even revealing the nature of their disease, the researchers never allowed the subjects to understand and participate in their treatment.

Again, what is the legacy of Tuskegee? Many have claimed that its effects are far-reaching and still present today (Jackson 2018).[4] One writer put it this way:

> The most enduring legacy of the Tuskegee Syphilis Study is its repercussions in the African American community … The study laid the foundations for African Americans' continued

---

[2] U.S. Public Health Service Syphilis Study at Tuskegee. Center for Disease Control and Prevention Web site. www.cdc.gov/tuskegee/index.html. Published 2020. Accessed 2/15/2021.

[3] U.S. Public Health Service Syphilis Study at Tuskegee. Center for Disease Control and Prevention Web site. www.cdc.gov/tuskegee/index.html. Published 2020. Accessed 2/15/2021.

[4] Tuskegee Syphilis Study. In: Lock S, Last JM, and Dunea G (eds.), *The Oxford Illustrated Companion to Medicine*, Oxford University Press, 2001.

distrust of the medical establishment, especially public health programs and a fear of vaccinations (Heintzelman 2003).

No apology or case analysis can undo these harms. We can only hope to learn from this bitter episode so that it may never happen again.

The Willowbrook Hepatitis Studies

It was the fall of 1975. Writing now as one of the authors of this book (DMS), I was in my second year of medical school. At 8:00 in the morning, I sat down in my usual seat towards the back of the auditorium as a medical specialist began a lecture on hepatitis. I was busily taking notes as the lecturer described the two major viral variants known at that time: hepatitis A and hepatitis B. And then he said something like this: "As we learned from the Willowbrook studies, hepatitis A is a relatively mild illness, with an incubation period of 3–4 weeks, and symptoms of fatigue, right-sided abdominal pain, and jaundice…" To my surprise, five students in the crowd suddenly got up from their seats and walked out of the room. I thought this was a bit rude, but I was too busy to think much more about it.

That is not the end of the story. The very next morning, the same lecturer came back in front of the class. He had not been scheduled to speak that morning. Accompanied by the Dean of Student Affairs, he formally apologized for referencing the Willowbrook studies, saying that it was unethical of him, promising it would never happen again. I had the impression that his apology was a condition for his continued employment as a member of the faculty. What was going on here?

Willowbrook State School, in Staten Island, New York, was an institution for intellectually disabled children from 1947 to 1987. At one time, it had over six thousand residents, even though the buildings were designed to accommodate just four thousand. During this era, many parents believed that institutionalizing disabled children was preferable to raising them at home, in the mistaken belief that a center such as Willowbrook could provide more services. It turns out that just the opposite was true: the conditions were deplorable, with many children neglected and malnourished. There certainly was no "education" taking place. One account puts it this way:

> Like many of the state-run institutions of its time, Willowbrook was a setting that purported to improve children's lives, but instead saw their overall quality of life and health deteriorate. (Scanlon 2007)

These conditions led to a high rate of infection with hepatitis A among the children at Willowbrook, which gave impetus for an experiment on its residents. Led by Dr. Saul Krugman of New York University, a research team deliberately infected some of the children on the thin rationale that they were likely to contract the disease anyway. One purpose was to study the natural history of hepatitis A, a goal disturbingly similar to that of the Tuskegee Syphilis Study (Rothman 1982). Lysaught describes the moral issues cogently:

At stake was almost every major issue in research ethics: that children were intentionally infected with a pathogen known for high risk of morbidity and mortality; that the children selected were mentally handicapped; that the parental permission letter did not make clear that the children would be intentionally infected; that, due to overcrowding and long waiting lists, parents might have been unduly influenced to enroll their children, as rooms in the experimental wing were readily available; and that many parents who sought placement at the school generally had insufficient resources to keep their children at home or to place them elsewhere (Lysaught 2009).

The public's pre-existing disgust about conditions at Willowbrook guaranteed a huge public outcry against the hepatitis experiments, even though Krugman was never officially censured for his work. The research, in fact, had tangible results. In essence, it led to a better understanding of hepatitis A as a disease and to the successful development of gamma globulin injections as a form of passive immunization against it (Hilleman 2010). Those pragmatic benefits notwithstanding, the Willowbrook studies were cited even in Henry Beecher's landmark article, "Ethics and Clinical Research," where he lists 22 egregious examples of unethical research studies in the early twentieth century; The Willowbrook study was number 16 on his list (Beecher 1966).

Therefore, the use of ethically questionable research by a clinical lecturer is a problem, and I now understand why my classmates left the lecture hall on that day so long ago in medical school. In Beecher's words, "An experiment is ethical or not at its inception; it does not become ethical *post hoc* – ends do not justify means" (Beecher 1966). Unethical methods will forever corrupt and taint the validity of any study on human subjects, no matter how clinically meaningful its results may appear to be.

Human Subjects Research in the Belmont Era

We could cite many examples of abuses of the rights of research subjects. Nonetheless, these examples, along with our earlier discussion on medical experiments performed by the Nazis, will suffice for our purposes. As discussed in the previous chapter, a team of American judges deliberated and passed judgment on the 23 medical doctors and scientists that participated in the experiments on prisoners in the concentration camps in World War II. During the Doctors' Trial, the assembled jurists also formulated a standard for the ethical treatment of human subjects called the Nuremberg Code (Shuster 1998).[5] This code, endorsed by the World Medical Association, was reconfigured as the Helsinki Declaration in 1964, and subsequently amended several times.[6]

---

[5] Code N. Trials of war criminals before the Nuremberg military tribunals under Control Council Law No. 10. *Washington, DC: US Government Printing Office*. 1949;2:181–182.
[6] World Medical Association Declaration of Helsinki: Ethical Principles for Medical Research Involving Human Subjects. *JAMA*. 2013;310(20):2191–2194.

Nonetheless, the revelations about the Tuskegee syphilis study in 1972 led to such widespread outrage that the U.S. Congress established the National Commission for the Protection of Human Subjects of Biomedical and Behavioral Research. The eleven members of this body spent four years to produce the comprehensive document we now know as the Belmont Report. Its content "reflects the committee's fears that research subjects would continue to be deceived, harmed, or otherwise exploited by investigators" (Friesen et al. 2017). The patient-centered principles of the Belmont Report have been enormously influential, and today govern well-established standards of consent and protection of vulnerable research populations. They are the basis of the modern Institutional Review Board (IRB), to be discussed later in this chapter.

Key Ethical Provisions of the Belmont Report

The Belmont Report codified a consensus that patient welfare should be the first ethical foundation of clinical research, not a pragmatic, utilitarian focus on the greatest benefit for the greatest number. The first part of the report makes a critical distinction between *clinical research* and *clinical practice*. Research is "an activity designed to test a hypothesis, permit conclusions to be drawn, and thereby to develop or contribute to generalizable knowledge." Clinical practice, on the other hand, can be defined as "interventions that are designed solely to enhance the well-being of an individual patient or client and that have a reasonable expectation of success."[7] In drawing a sharp line between research and practice, the goal of Belmont was to ensure that human rights were adequately protected and that the excesses of Tuskegee, Willowbrook, and a myriad of other unethical, discriminatory, and harmful projects would never happen again.

The Belmont Report goes on to articulate three foundational ethical principles "generally accepted in our cultural tradition:" *respect for persons, beneficence, and justice.*[8] Though framed in a slightly different way, we are already familiar with these ideas, which emerge from the Hippocratic Oath and medical principlism (Chap. 4).

Respect for persons relies on ancient ideas of human value that emerge from Judeo-Christian precepts, as well as Aristotelian ideas that revolve around human agency and worth (see Chap. 3). This precept implies the right of every person to

---

[7] The Belmont report: Ethical principles and guidelines for the protection of human subjects of research. Superintendent of Documents. National Commission for the Protection of Human Subjects of Biomedical and Behavioral Research Web site. www.hhs.gov/ohrp/regulations-and-policy/belmont-report/read-the-belmont-report/index.html. Published 2018. Accessed 2/15/2021.

[8] The Belmont report: Ethical principles and guidelines for the protection of human subjects of research. Superintendent of Documents. National Commission for the Protection of Human Subjects of Biomedical and Behavioral Research Web site. www.hhs.gov/ohrp/regulations-and-policy/belmont-report/read-the-belmont-report/index.html. Published 2018. Accessed 2/15/2021.

make his or her own decisions about medical treatments or participation in research. Medical principlism, of course, would express this as *autonomy*, which undergirds un-coerced informed consent (Adashi et al. 2018).

The second major Belmont principle is *beneficence*. We earlier expressed this idea from the twin Hippocratic ideas of beneficence (intending the best interests of a patient) and non-maleficence (avoiding harm), perhaps a more detailed way to encompass Belmont's overall intent (Beauchamp 2010).

Finally, Belmont focuses on *justice*. Related to the Hippocratic principle of distributive justice, it implies that clinical research on human subjects should be inclusive and allow for full participation by all segments of society. Furthermore, the expected benefits must help all members of the population (Adashi et al. 2018).

Though never discussed in the Belmont Report itself, a beneficial idea emerges by combining the first and second parts, i.e., the research/practice distinction and the ethical principles. This idea is the concept of *clinical equipoise*. Consider the following rhetorical example. A new cancer drug emerges on the market to treat lung cancer. It has obtained FDA-approval for this indication and has been shown to improve patient survival. Given this knowledge, is it ethical to perform a randomized clinical trial where the drug is given to an "experimental" group for this indication, but not to a control group? Obviously not, since the drug has already undergone experimental testing – it has passed from the research realm into the realm of clinical practice. A randomized clinical trial can only be ethical when a true state of *equipoise* exists, where the researchers do not know if the drug is helpful or not. One writer put it this way:

> Clinical equipoise is the assumption that there is not one 'better' intervention present (for either the control or experimental group) during the design of a randomized controlled trial. A true state of equipoise exists when one has no good basis for a choice between two or more care options. Clinical equipoise has also been called an honest null hypothesis and/or a state of uncertainty (Freedman 1987).

Here is another description of the same idea:

> Equipoise refers to a state of regarding two treatments as an equal bet in prospect… If there is 'absolute' uncertainty, the decision-makers are 'agnostic' or in 'equipoise,' that is, the 'mean benefits in prospect from both treatments are equal…' (Edwards et al. 1998)

This concept of clinical uncertainty or equipoise is a major benchmark for understanding research ethics. The more we are certain of the potential benefits of a clinical intervention, the less we are justified in denying that treatment to one group of patients in the name of research.

The Institutional Review Board

In 1981, based on the Belmont Report, the Department of Health and Human Services (DHHS) and the Food and Drug Administration established regulatory guidelines for research on human subjects that are still in place to this day. Now regulated by the Office of Human Research Protections (OHRP, under DHHS), this is all codified as Part 46 (Protection of Human Subjects) of Title 45 of the Code of Federal Regulations, also called the Revised Common Rule.[9] The Revised Common Rule has led to the development of the modern Institutional Review Board (IRB).

Every institution in the U.S. that performs research on human subjects must have an IRB committee in place or must coordinate with an outside IRB. Some IRBs are small and not officially registered with the OHRP in Washington, DC. In contrast, others are larger and have official recognition (called a *Federal Wide Assurance*), but all IRBs must comply with federal standards for human research protection. A full discussion of the IRB review process lies outside the scope of this text, but a summary will suffice for our purposes.

The primary goal of an IRB is to protect the interests of subjects enrolled in clinical trials. In reviewing a proposed research project, no study may cause intentional or preventable injury, loss of function, or the undermining of personal dignity. The potential benefits of participation for research participants must outweigh the risks. There must also be no foreseeable violation of privacy or exposure to embarrassment or legal liability. Justice requires that all participants be treated in the same manner, regardless of gender, sexual orientation, ethnicity, socio-economic status, religion, or other non-relevant factors. All human studies must be reviewed and approved by an IRB before any research activities can begin.

Finally, certain vulnerable special groups require extra protection and consideration. These include pregnant women, fetuses, children, prisoners, and those with physical and mental disabilities. Research involving such groups will require particular scrutiny. For all research participants in the healthcare arena, strict adherence to principles of confidentiality of private information must be maintained to comply with HIPAA regulations (Health Insurance Portability and Accountability Act of 1996) (Cargill et al. 2016).[10, 11]

In summary, this chapter has examined the sordid and disturbing Tuskegee Syphilis project, which persisted as a research initiative for more than 40 years. We then discussed the Willowbrook hepatitis study, where disabled children received injections of the hepatitis A virus. Both of these examples, plus many not explicitly mentioned, have left behind a negative legacy of research on vulnerable populations, with many repercussions still present today.

To protect human research subjects from such abuses, the Belmont Report of 1979 clearly articulates the difference between clinical research and clinical treatments. It specifies the three primary principles of respect for persons, beneficence, and justice. One practical implication of Belmont is the concept of clinical

---

[9] Part 46 – Protection of Human Subjects. Code of Federal Regulations Web site. www.ecfr.gov/cgi-bin/text-idx?m=04&d=01&y=2018&cd=20190401&submit=GO&SID=46df0dc4d480c687973240ccd0f5fd6a&node=pt45.1.46&pd=20180719. Published 2018. Accessed 2/15/2021.

[10] Adams LA, Callahan T. Research Ethics. University of Washington. https://depts.washington.edu/bhdept/ethics-medicine/bioethics-topics/detail/77. Published 2013. Accessed 2/15/2021.

[11] Vulnerable Subjects. Institutional Review Board for Health Sciences Research Web site. https://research.virginia.edu/irb-hsr/vulnerable-subjects. Published 2020. Accessed 2/15/2021.

equipoise, the idea that true therapeutic uncertainty must be present for a randomized, placebo-controlled clinical trial to be ethical.

Finally, Institutional Review Boards have arisen during the Belmont era: oversight committees at every hospital, academic center, or clinic that performs human subject research. Such committees help to ensure that human research participants remain safe.

So far, we have completed three chapters on major historical events relating to medical ethics. The next chapter will conclude this section on ethical foundations with a discussion of pharmacy professionalism.

Key Terms

- Doctor's Trial
- Nuremberg Code, Helsinki Declaration
- Belmont report
- Clinical equipoise
- Institutional Review Board

Questions for Review and Discussion

1. Summarize the key ethical violations of the Tuskegee Syphilis program in light of medical principlism.
2. What was ethically wrong with the Willowbrook study? Why does it remain problematic to cite its data in lectures about hepatitis today?
3. Describe the key elements of the Belmont Report. What three ethical principles form its foundations?
4. Define clinical equipoise and explain how this concept helps us to understand the ethics of human clinical research trials.
5. What is an Institutional Review Board? List the major vulnerable population groups that require special protections in research.

Further Reading

James H. Jones, *Bad Blood: the Tuskegee Syphilis Experiment.* Free Press, 1981. This is the book that started it all, a painstakingly researched yet very readable account that lays out in chilling detail the abuses of Tuskegee.

Rebecca Skloot, *The Immortal Life of Henrietta Lacks*, Broadway Paperbacks, 2011. Anyone who has every worked with *HeLa* cells in tissue culture will want to read this account of an African-American woman in Baltimore, whose cervical biopsy tissue was used for research without her permission. The author's meticulous investigation for over a decade finally brought this abuse to public light. The book has won a host of prestigious awards.

References

Adashi EY, Walters LB, Menikoff JA. The Belmont report at 40: reckoning with time. Am. J. Public Health. 2018;108(10):1345–8.
Beauchamp TL. Standing on principles: collected essays. Oxford: Oxford University Press; 2010.
Beecher HK. Ethics and clinical research. N. Engl. J. Med. 1966;274(24):1354–60.
Brandt AM. Racism and research: the case of the Tuskegee Syphilis Study. Hast. Cent. Rep. 1978;8(6):21–9.
Brown DL, Frank JE. Diagnosis and management of syphilis. Am. Fam. Phys. 2003;68(2):283–90.
Cargill SS, DeBruin D, Eder M, et al. Community-engaged research ethics review: exploring flexibility in federal regulations. IRB. 2016;38(3):11–9.
Edwards SJ, Lilford RJ, Braunholtz DA, Jackson JC, Hewison J, Thornton J. Ethical issues in the design and conduct of randomised controlled trials. Health Technol. Assess. (Winch., Eng.). 1998;2(15):i.
Freedman B. Clinical equipoise and the ethics of clinical research. N. Engl. J. Med. 1987;317:141–5.
Friesen P, Kearns L, Redman B, Caplan AL. Rethinking the Belmont report? Am. J. Bioethics. 2017;17(7):15–21.
Heintzelman CA. The Tuskegee syphilis study and its implications for the 21st century. New Soc. Worker. 2003;10(4):4–5.
Hilleman M. Vaccine research using children. In: Norkin L, editor. Virology: molecular biology and pathogenesis. Washington, DC: ASM Press; 2010.
Jackson K. A legacy of shame. Today's Geriatr. Med. 2018;11(4):4–4.
Jones JH. Bad blood: the Tuskegee syphilis experiment. New York: Free Press – Collier Macmillan Publishers; 1981.
Lysaught MT. Docile bodies: transnational research ethics as biopolitics. J. Med. Philos. 2009;34(4):384–408.
Rothman DJ. Were Tuskegee & Willowbrook 'studies in nature?'. Hast. Cent. Rep. 1982;12(2):5–7.
Scanlon E. Willowbrook state school. In: The Praeger handbook of special education. 2007. pp 10–12.
Shuster E. The nuremberg code: Hippocratic ethics and human rights. Lancet. 1998;351(9107):974–7.

Chapter 7
Pharmacy Professionalism

The Ethics Code of Pharmacists

The first six chapters of our text laid the ethical foundations for modern pharmacy practice. We considered fundamental ethical theories as well as competing ideas of human value. We then reviewed clinical ethics in historical context, beginning with the ancient traditions of the Hippocratic Oath, to which we added the eighteenth century principle of autonomy. Our historical analysis then examined some terrible departures from these standards in the twentieth century, including the flawed pseudo-science of eugenics and the horrific excesses of Nazi medicine. We concluded the opening chapters by looking at the Tuskegee syphilis trials and the Willowbrook hepatitis study, abuses that both helped to provide the impetus for the 1979 Belmont Code and the development of modern institutional review boards. We now conclude this foundational section with the tenets of pharmacy professionalism, as exemplified in the Ethics Code of the American Pharmacists Association (APhA). Later chapters will then lead us into practical matters in the form of issues and cases.

As we saw in the opening chapter of this text, pharmacy ethics as a separate discipline is relatively new, beginning in the latter half of the twentieth century. The increasing complexity of pharmacy practice and the changing role of pharmacists as members of the healthcare team have led to the need for a greater emphasis on ethics. Therefore, in 1989, the Joint Commission of Pharmacy Practitioners called on the American Pharmaceutical Association (which later became APhA) to adopt a new ethics standard, formalized in 1994, known today as the Code of Ethics for Pharmacists (Buerki et al. 2013).

APhA is the largest professional organization of pharmacists in the U.S. and is active in education, leadership development, and advocacy for the pharmacy profession (About APhA 2021). Its Code of Ethics is well-known to students training to

---

This chapter features an additional author, PharmD candidate Jonathan Williams (class of 2021).

become pharmacists and is often recited at "white coat" ceremonies where students begin their graduate professional education. Its eight principles are as follows:

I. A pharmacist respects the covenantal relationship between the patient and pharmacist.
II. A pharmacist promotes the good of every patient in a caring, compassionate, and confidential manner.
III. A pharmacist respects the autonomy and dignity of each patient.
IV. A pharmacist acts with honesty and integrity in professional relationships.
V. A pharmacist maintains professional competence.
VI. A pharmacist respects the values and abilities of colleagues and other health professionals.
VII. A pharmacist serves individual, community, and societal needs.
VIII. A pharmacist seeks justice in the distribution of health resources (APhA 2019).

Right from the beginning, the Code reflects the influence of the Hippocratic Oath, which begins: "…I will fulfill according to my ability and judgment, this Oath and this covenant" (Edelstein 1943). Note the first tenet of the Code of Ethics for Pharmacists: *A pharmacist respects the covenantal relationship between the patient and pharmacist*. As we discussed in Chap. 4, the relationship of the healthcare professional to the patient is more than a mere contract. The word "covenant" implies a deeply binding obligation, often described in our modern context as a "fiduciary relationship." The term *fiduciary* is a legal term based on trust. A fiduciary is "someone who has undertaken to act for or on behalf of another in a particular matter in circumstances which give rise to a relationship of trust and confidence. The distinguishing obligation of a fiduciary is the obligation of loyalty."[1] As one legal scholar has put it, fiduciaries must conduct themselves "at a level higher than that trodden by the crowd" (Conaglen 2010). In current usage, therefore, the word fiduciary has applications primarily among bankers, lawyers, and financial counselors. However, many ethicists have suggested that this is also the most descriptive term for the mutual, deeply binding obligations of the medical professional, whether referring to a physician, a nurse, or a pharmacist. This principle undergirds all of the remaining tenets of the Code of Ethics.

As an example, suppose that a patient discloses to you that she is trying non-medical marijuana (in a state where this is not legal) for a chronic pain issue. As a responsible and trustworthy pharmacist, you note this within your in-house system for medically relevant follow-up. Nonetheless, you do not report this technically illegal activity to authorities or disclose it to her family, friends, or anyone else without a sufficient reason to know. You keep the patient safe from clinical complications while protecting her confidentiality. Such a wise approach reflects a deep relationship of trust and confidence.

The second ethical principle of the Code: *A pharmacist promotes the good of every patient in a caring, compassionate, and confidential manner*. Note that this is

---

[1] *MOTHEW v. BRISTOL and WEST BUILDING SOCIETY*, (EWCA Civ 533 1996).

the idea of *beneficence*, which we earlier described as part of the Hippocratic tradition. The Code for pharmacists adds an attitude, that of compassion, and the concept of confidentiality, also explicitly mandated in the ancient Oath. This tenet also implies *non-maleficence*, the Hippocratic duty to minimize any harm to the patient.

Consider an elderly patient with early dementia, somewhat forgetful though still living on his own. As a compassionate pharmacist, you should be considerate and patient with this gentleman, setting aside time to carefully counsel him each time he comes in to pick up his medications. Though it is tempting to enlist other family members in his care, you should follow the patient's wishes as to what others should know, as far as possible.

The third component: *A pharmacist respects the autonomy and dignity of each patient*. As we mentioned previously, autonomy was absent from the Hippocratic Oath; instead, it comes from a much later period in history (the eighteenth century Enlightenment) under Immanuel Kant (Johnson and Cureton 2016). A pharmacist respects patients' autonomy by allowing them and their representatives to make their own informed healthcare decisions, as tailored by their personal values (Autonomy 2020; Varelius 2006). This egalitarian principle directly leads to informed consent and patient-centered care.

This part of the Code is sometimes tough to follow. Imagine that your patient has a different opinion on the adverse effects of a prescribed medication than the medical team. He is experiencing bothersome ankle edema from taking amlodipine for hypertension (Vukadinovic et al. 2019). If his blood pressure is only moderately elevated (say, 140/88), he may reasonably view the side effect as worse than his risk of a heart attack or stroke. You should, of course, educate him on his options, including having his physician add a diuretic for the edema. Nonetheless, you should respect his decision even if he disagrees with your advice.

The fourth and fifth statements of the Code of Ethics go hand in hand: *A pharmacist acts with honesty and integrity in professional relationships*, and *a pharmacist maintains professional competence*. These more inward-focused values take professionalism beyond the realm of mere good intentions; they refer to personal character and technical expertise, without which no pharmacist can succeed.

In a profession with such a high reputation for trust, pharmacists need to avoid temptations to act dishonestly for personal gain. Unfortunately, violations abound. It is all too common to see the practice of filling prescriptions with generics, yet billing for brand name medications, increasing costs to the patient and their insurers. Of course, stealing controlled substances, either for personal abuse or diversion for money, also occurs far too often and is one of the most common temptations pharmacists face (Combs 2008). In light of these dangers, pharmacists must have a high degree of integrity in practice.

Yet honesty and integrity must also be paired with technical competence. As pharmacy professionals age, they may easily become complacent and fail to stay current. They may "go through the motions" of continuing education without really paying attention, and patients may suffer. Imagine a pharmacist who fails to understand basic pharmacogenomics. If the patient is a known CYP 26D ultra-rapid metabolizer, a standard dose of codeine may be dangerous. In essence, the patient may activate more of the prodrug (codeine) into its active metabolite (morphine), leading to respiratory depression (Malan et al. 2019; Horn and Hansten 2008). Technical competence means keeping abreast of newer findings to avoid such complications.

The sixth component of the Code of Ethics is as follows: *A pharmacist respects the values and abilities of colleagues and other health professionals*. This idea is a valuable reminder that pharmacists are members of a specialty *guild* and that they owe their clinical expertise to the colleagues and professors who have gone before them. Even the ancient Oath of Hippocrates acknowledges this debt: "To hold him who has taught me this art as equal to my parents…" (Edelstein 1943).

Let's be clear on this point: modern healthcare is all about the healthcare *team*, but it doesn't always function that way. In a perfect world, physicians and nurse practitioners would actively seek input from pharmacists, who, after all, are the medication experts. But all too often competition and professional jealousies get in the way. The resulting poor communication may lead to less than optimal patient outcomes. Fairly often, a pharmacist may view a doctor's choice of medication as sub-optimal, leading to possible side effects, higher costs, or difficulty taking that dosage form. As an example, consider a prescription for metoprolol tartrate in a patient with both asthma and heart failure. Since metoprolol is somewhat beta-1 selective, the doctor may assume that this is acceptable to take along with their beta-2 agonist albuterol rescue inhaler. The pharmacist should know that higher doses of selective beta-1 blockers begin to block many beta-2 receptors, leading to a decreased efficacy of albuterol. The pharmacist may, therefore, strongly prefer a moderate dose of metoprolol succinate, an extended-release formulation that does not have as high a peak (Lopressor 2020; Toprol 2020). However, other factors may lead to a preference for the tartrate salt formulation, such as lower cost and that particular patient's rare need for the inhaler. The pharmacist should respect the doctor's choice if there are adequate benefits and not much potential for harm. At the same time, this must be balanced against what is clinically optimal. Ideally, a polite call to the prescriber should resolve the matter, but respect for one another is the key. Communication and trust are essential for a highly functional team.

The last two statements (the seventh and eighth) in the Code of Ethics for Pharmacists are these: *A pharmacist serves individual, community, and societal needs*, and A *pharmacist seeks justice in the distribution of health resources*. These tenets reflect the modern context of health care as a limited resource. With the aging of the U.S. population, it is becoming increasingly important for federal and state governments to control health care costs, leading to a greater reliance on managed care. In the continual tug-of-war between private and public sources of payment, there may be an unequal distribution of benefits. A pharmacist must recognize this inherent conflict between the needs of the community and those of individual patients and must seek distributive justice for all. The astute reader will recall from Chap. 4 that *distributive justice* is a Hippocratic principle that refers to the fair and equitable distribution of the benefits of healthcare (Beauchamp and Childress 2019). In our modern society, this means that there can be no barriers for patients to access clinical treatments, regardless of age, gender, social class, ethnicity, sexual orientation, ability to pay, religion, disability, or any other medically non-relevant trait. A pharmacist must seek to mitigate any barriers that persist.

Increasingly, therefore, a pharmacist must look outward to the rest of society to completely fulfill his or her ethical obligations. Nevertheless, embedded within this duty lies an important caveat and warning: *always look inward first*. The foundation of the Code of Ethics, and of the Hippocratic tradition from which it derives, is that of a *covenant*, the fiduciary obligation of the healthcare professional to the individual patient. A pharmacist, or any other healthcare professional, must not be confused by broader societal concerns in the middle of taking care of a particular *someone. One cannot make policy at the bedside*.[2]

Virtue as the Foundation of Pharmacy Professionalism

We began with the Code of Ethics for Pharmacists as a basic framework for how to conduct the practice of pharmacy ethically and professionally. But what drives all of this? What is the motivation to become a pharmacist and excel at this unique patient care role? We would propose that the underlying drive in pharmacy is *virtue*. If finances were the sole motivation for becoming a pharmacist, one might be better served by heading into another field. The fact is that, as clinicians, we all care deeply for our patients and seek to act according to their best interests, completing and improving their care whenever we can. Virtue is the thread that ties us all together as an ideal foundation for moral reasoning. As an ethical theory, as discussed in Chap. 2, it centers on strength of character. Virtuous character and a motivation to do good toward others are the very essence of health care, shared by pharmacy and other clinical disciplines.

Nevertheless, we must acknowledge the limitations of virtue as the primary ethical stance for pharmacy practice. Not everyone will have an identical list of traits that all can agree are virtuous. Furthermore, it is not always clear which action is the most virtuous, even given an established list of virtuous traits. Virtue might be enough for simple cases, such as saving a patient's life if you know you are able. However, in complex matters, a virtuous intent alone may be insufficient.

And so, we propose that virtue be used as a central *motivation* for ethical decision making, but combined with other ethical theories for more complex decision-making. Other major ethical theories (from Chap. 2) include *utilitarianism, Kantian ethics, medical principalism, natural law,* and *divine command theory*. We suggest that all pharmacists have a basic understanding of each of these theories and be able to apply them in practice. Pharmacists should also be able to understand the argument when someone else speaks from one or more of these ethical frames. Even if we disagree with another's conclusion, we should still be able to see where they come from, and balance our consideration of the issue around multiple perspectives.

---

[2] The lead author is indebted to ethicist Dr. Robert D. Orr for this insight.

We should also be sensitive to the views of our patients, their families, and our fellow clinical professionals in coming to an ethical decision. Humility, in such cases, is itself an important virtue. We will continue to develop this idea of combining virtue ethics with other ethical theories in later chapters. For now, we will focus on the application of virtue ethics to health and wellness promotion and professionalism in pharmacy practice.

Health and Wellness Promotion

Community-based retail pharmacists are more accessible than most other healthcare professionals. Given time and motivation, they can make a significant impact on the health and quality of life of their patients. Such "wellness promotion" goes beyond the mere filling of prescriptions. One group of pharmacy educators put it this way:

> With community pharmacists expanding their focus beyond dispensing to the broader goal of helping patients achieve health and wellness, they are positioned to improve community health. Community pharmacies are easily accessible and most have extended hours. Further, patient acceptance of pharmacy-based wellness and screening services has been demonstrated in previous studies, including the willingness to pay out of pocket for such services. (DiDonato et al. 2013)

There is a growing awareness that pharmacists add significantly to the effectiveness of primary care initiatives. Manolakis and Skelton point out that pharmacy practice is a crucial component to "health promotion, disease prevention, health maintenance, counseling, and patient education." This conclusion is especially true since "medications play a critical role in primary care, and their effective management is essential to patient safety and quality care" (Manolakis and Skelton 2010).

What kind of activities are we talking about? Screening and patient education are two of the most important. Pharmacists have the training to assess health issues in their patients, including screening for chronic disease risk factors. Such screening may include weight, blood pressure, and lipid and glucose levels, to name just a few. Of course, mere information is not sufficient. Assessing risk factors must be combined with educational interventions, such as diet counseling for weight loss and blood pressure control, smoking cessation advice, input on physical activity, and recommendations for vaccinations (Mossman 2006).

As far as medications are concerned, medication therapy management (MTM) is a growing part of every pharmacist's role. This phrase is a blanket term for "patient-focused services aimed at improving therapeutic outcomes" (Group and Burns 2005). The American College of Clinical Pharmacy provides a more comprehensive definition:

> [MTM is] a review of all medications prescribed by all prescribers providing care to the patient, and any over-the-counter and herbal products the patient may be taking to identify and address medication problems. Problems may include medications not being used correctly, duplication of medications, unnecessary medications, and the need for medication(s)

for an untreated or inappropriately managed condition. (American College of Clinical Pharmacy 2020)

As stated earlier, all of these endeavors are important elements to improve the overall impact of primary care.

Everyday Professionalism

On the job, busy community or hospital-based pharmacists don't always deal with controversial matters that require deep reflection. Yet basic courtesy and a desire to be helpful are also important elements of ethical duty, and the Code of Ethics speaks to these as well. One of our former students offers this list of daily small ethical challenges, and appended relevant sections of the Code:

- Dealing with difficult patients, respectfully and calmly (I, II, III).
- Handling questionable and perhaps even forged prescriptions (IV, V).
- Knowing your limits and calling another colleague when a situation is beyond what you are prepared for (II, IV, VI, VII).
- Balancing a desire for patients to have accurate health information with sensitivity to their culture and beliefs (II, III).
- Always offering to counsel every patient with a new prescription. Be sure to provide a private environment for this, to protect confidentiality (II, III).
- Handling calls to prescribers with tact. They may be frustrated by frequent calls, business issues, any disagreements with their therapy selection, and other irritations. Always put the patient first and be considerate of the prescriber during these potentially difficult conversations. Don't take anything negative personally, and be willing to let minor things go (IV, V, VI).[3]

How we handle each of these small daily interactions is a reflection of our own standards of ethics and indeed our virtuous character. The Code and the virtues that underlie it constitute sound advice. Following these standards will require humility and a realization that none of us are perfect.

Recent Developments

The past half-century has seen a significant shift in the focus of the pharmacy profession. Two significant movements have affected the need for a greater emphasis on ethics: (1) a change from being a product-distribution centered profession to a more clinical, patient-centered profession, and (2) the shift from being mostly

---

[3] These insights specifically offered by PharmD candidate Jonathan Williams (class of 2021), who has also co-authored this chapter.

self-employed to a profession of almost entirely employee pharmacists. The rise of large chain pharmacies has driven the latter, along with changes in payment mechanisms for prescriptions and the increasing utilization of pharmacists in healthcare systems. Since 2010, the number of independent community-based ("retail") pharmacies has decreased from 23,064 to 21,767 in 2017, a 5.62% decrease (National Community Pharmacists Association 2019).

More recently, collaborative practice agreements (CPAs) have allowed physicians and pharmacists to share responsibilities and treatment authority for many disease entities (Centers for Disease Control and Prevention 2017). Utilizing CPAs has allowed pharmacists to obtain prescribing rights, the ability to order laboratory and other diagnostics, and patient assessment privileges, all of which are necessary for the proper management and monitoring of medication therapy. The Veterans Health Administration, for example, is well known for providing these types of services. Furthermore, several states have granted pharmacists provider status to enable billing for clinical services (Ross 2015; Balick 2017). Within many health systems, pharmacists can add, remove, or modify medication therapy and order laboratory tests under hospital protocols. The increased focus of pharmacists on patient-centered and clinical practices will also increase their encounters with ethical dilemmas.

Pharmacy Professionalism and the Law

This text does not focus a great deal on pharmacy laws and regulations. However, it will be instructive to discuss how pharmacy professionalism interacts with the law, especially concerning the possible delegation of a pharmacist's duties. We begin with two cautionary examples.

Eric Cropp was a pharmacist at Rainbow Babies and Children's Hospital in Cleveland, Ohio (Vivian 2009). On February 26, 2006, Cropp received an order for etoposide in 0.9% saline solution for 2-year-old Emily Jerry. This treatment was to be her final dose of chemotherapy. Instead of using a premixed bag of normal saline, pharmacy technician Katie Dudash, supervised by Cropp, chose to prepare the dose using concentrated 23.4% sodium chloride, which resulted in a hypertonic saline solution that was likely above 20% (Berger 2018; Emily Jerry Foundation 2020). Cropp failed to detect the error, which resulted in Emily's death, devastating the Jerry family.

The hospital terminated Eric Cropp, and the Ohio Board of Pharmacy revoked his license. In 2009, Cropp was convicted of involuntary manslaughter and spent 6 months in prison and 6 months in house arrest. He incurred a $5000 fine plus court costs and performed 400 hours of community service. The incident prompted the passage of "Emily's Law" in Ohio, which requires pharmacy technicians to be at least 18 years old and to pass a competency test and a criminal background check (Vivian 2009; Berger 2018; Emily Jerry Foundation 2020).

Our second example is from Cedars-Sinai Medical Center in Los Angeles, where, in 2014, a pharmacy technician took heparin from the pharmacy's supply without carefully checking the concentration. Cedars-Sinai at the time had a "tech-check-tech" policy, where the heparin dose was supposed to have been verified by a second technician, which didn't occur. The nurses who administered the doses also failed to recognize the error (Ornstein 2014; Cedars-Sinai Newsroom 2007).

The heparin was intended as a simple flush to prevent blood clot blockage of intravenous catheters. As a result of the mistake, the dose administered to three children had a concentration of 10,000 units per milliliter, rather than 10 units per milliliter. Despite this error, all three infants eventually recovered and were discharged (Ornstein 2014; Cedars-Sinai Newsroom 2007). The hospital was fined by the California Department of Public Health, and the case prompted the Joint Commission for the Accreditation of Health Organizations to establish National Patient Safety Goals related to anticoagulation (*Drug Topics* 2008).

What do these two cases have in common? They both represent inadequate supervision of pharmacy technical personnel, one by negligence and the other by system design. In both instances, inadequate supervision led to avoidable errors, with a tragic result in one case. Some legal terminology is instructive here. In the example of Eric Cropp, the *strict liability* standard applies. This phrase means that he alone bears the responsibility of the error, rather than the technician (Sveska 1993). It was his duty to supervise the technician, who suffered minimal penalties as a result of her mistake (Vivian 2009). Note that strict liability is not the same as the *corresponding responsibility* pharmacists have with prescribers for controlled substance orders. In corresponding responsibility, the pharmacist shares equally with the prescriber the culpability for filling inappropriate or illegal prescriptions (Williams et al. 2017).

In the Cedars-Sinai case, the step of a pharmacist checking the dose was bypassed in the system by design. We recognize that tech-check-tech has many strong proponents and has increased in usage since the Cedars-Sinai incident (Tarver et al. 2017). Nonetheless, in both cases, technical functions, admixture, and dispensing, were *delegated* to pharmacy technicians. But certain professional duties should not be delegated.

The legal concept of *respondeat superior* is helpful to understand this. This phrase is a legal doctrine that translates, "let the master answer" (Respondeat Superior 2020). Usually applied in civil law, it simply means that there is a person or entity that has ultimate responsibility for an injury or harm. In practice, this means that the fiduciary duty of pharmacists to their patients *cannot be delegated* to technical or clerical personnel. The ethical obligations of beneficence and nonmaleficence belong to the pharmacist alone and cannot be transferred to a technician or anyone else. This commitment is the high standard of professionalism.

Conclusion

Though much more could be said, this chapter has introduced the major tenets of pharmacy professionalism. We began with the Code of Ethics for Pharmacists, pointing out its affinities with the Hippocratic Oath and with medical principlism. We then briefly examined the idea that virtue ethics might provide the foundation for professional practice. Finally, we looked at some specific ways to carry out these precepts in daily practice and considered how pharmacy professionalism interacts with the law.

This discussion concludes Section I of this text, which considers the foundations for pharmacy ethics. We now turn to Section II: Issues and Cases, beginning in Chap. 8 with reproductive ethics.

Key Terms

- Covenant
- Fiduciary relationship
- Beneficence, non-maleficence
- Distributive justice
- Autonomy
- Virtue ethics
- Medication therapy management
- Collaborative practice agreements
- Strict liability
- Corresponding responsibility
- *Respondeat superior*

Questions for Review

1. Memorize and be able to describe the eight tenets of the Code of Ethics for Pharmacists.
2. In what way do the last two tenets of the Code of Ethics differ from the others? How is this difference significant ethically?
3. Why do the authors recommend virtue ethics as the foundation for pharmacy professionalism?

For Further Reading

David Tipton, *Professionalism, Work, and Clinical Responsibility in Pharmacy*. Jones and Bartlett, 2013. This systematic text explores a pharmacist's role and responsibilities from the perspective of the American Board of Internal Medicine model of professionalism. It uses the ABIM's six attributes: altruism, accountability, duty, excellence, honor and integrity, and respect for others. The reader will find it instructive to compare this approach to our emphasis on virtue ethics.

References

About APhA. The American Pharmacists Association Web site. www.pharmacist.com/about-apha. Published 2021. Accessed 15 Feb 2021.

American College of Clinical Pharmacy. What is medication therapy management? American College of Clinical Pharmacy Web site. https://www.accp.com/. Published 2020. Accessed 15 Feb 2021.

APhA. APhA code of ethics. American Pharmacists Association Web site. www.pharmacist.com/code-ethics. Published 2019. Accessed 15 Feb 2021.

Autonomy. Oxford English dictionary. 2020. https://en.oxforddictionaries.com/definition/autonomy

Balick R. New Tennessee law formally recognizes pharmacists as providers. American Pharmacists Association. 2017. www.pharmacist.com/article/new-tennessee-law-formally-recognizes-pharmacists-providers. Published 14 Aug 2017.

Beauchamp TL, Childress JF. Principles of biomedical ethics. 8th ed. New York: Oxford University Press; 2019.

Berger K. Eric Cropp discusses medical error that sent him to prison. Pharmacy Times. 2018. www.pharmacytimes.com/contributor/karen-berger/2018/03/getting-to-know-the-caring-eric-cropp. Published 21 Mar 2018.

Buerki RA, Vottero LD, American Pharmacists Association. Pharmacy ethics: a foundation for professional practice. Washington, DC: American Pharmacists Association; 2013.

California hospital fined for heparin overdose. Drug Topics. 2008. www.drugtopics.com/modern-medicine-news/california-hospital-fined-heparin-overdose. Published 24 Mar 2008.

Cedars-Sinai Newsroom. Tech-Check-Tech: new California regulation will help prevent medication errors, free pharmacists for more direct patient care. Cedars-Sinai Newsroom Web site. www.cedars-sinai.org/newsroom/tech-check-tech-new-california-regulation-will-help-prevent-medication-errors-free-pharmacists-for-more-direct-patient-care/. Published 2007. Accessed 15 Feb 2021.

Centers for Disease Control and Prevention. Advancing team-based care through collaborative practice agreements a resource and implementation guide for adding pharmacists to the care team. Centers for Disease Control and Prevention Web site. www.cdc.gov/dhdsp/pubs/docs/CPA-Team-Based-Care.pdf. Published 2017. Accessed 15 Feb 2021.

Combs J. Incomprehensible demoralization: an addict pharmacist's journey to recovery. Xlibris; 2008.

Conaglen M. Fiduciary loyalty: protecting the due performance of non-fiduciary duties. Oxford/Portland: Hart Publishing; 2010.

DiDonato KL, May JR, Lindsey CC. Impact of wellness coaching and monitoring services provided in a community pharmacy. J Am Pharm Assoc. 2013;53(1):14–21.

Edelstein L. The Hippocratic oath, text, translation and interpretation. Baltimore: The Johns Hopkins Press; 1943.

Emily Jerry Foundation. Emily's story. Emily Jerry Foundation Web site. https://emilyjerryfoundation.org/emilys-story/. Published 2020. Accessed 15 Feb 2021.

Group TL, Burns A. Medication therapy management services: a critical review. J Am Pharm Assoc. 2005;45(5):580–7.

Horn JR, Hansten PD. Get to know an enzyme: CYP2D6. Pharmacy Times. www.pharmacytimes.com/publications/issue/2008/2008-07/2008-07-8624. Published 1 July 2008.

Johnson R, Cureton A. Kant's moral philosophy. Stanford encyclopedia of philosophy. 2016. https://plato.stanford.edu/entries/kant-moral/

Lopressor. Rx List. 2020. www.rxlist.com/toprol-xl-drug.htm#description

Malan L, Lundie M, Engler D. Oral opioid metabolism and pharmacogenetics. SA Pharm J. 2019;86(2):21–8.

Manolakis PG, Skelton JB. Pharmacists' contributions to primary care in the United States collaborating to address unmet patient care needs: the emerging role for pharmacists to address the shortage of primary care providers. Am J Pharm Educ. 2010;74(10):1–9.

Mossman J. The pharmacist's role in community wellness. US Pharm. 2006;1:43–50.

National Community Pharmacists Association. NCPA digest. National Community Pharmacists Association Web site. www.ncpa.co/pdf/digest/2019/2019-digest.pdf. Published 2019. Accessed 15 Feb 2021.

Ornstein C. Dennis Quaid files suit over drug mishap. Los Angeles Times. 2014. www.latimes.com/entertainment/gossip/la-me-quaid5dec05-story.html. Published 14 Sept 2014.

Respondeat Superior. Legal Information Institute. 2020. www.law.cornell.edu/wex/respondeat_superior

Ross M. Big win for pharmacist provider status in Washington State. Pharmacy Times. 2015. www.pharmacytimes.com/news/big-win-for-pharmacist-provider-status-in-washington-state. Published 11 May 2015.

Sveska KJ. Pharmacist liability. Am J Hosp Pharm. 1993;50(7):1429–36.

Tarver SA, Palacios J, Hall R, Franco-Martinez AC. Implementing a tech-check-tech program at a university health system. Hosp Pharm. 2017;52(4):280–5.

Toprol XL. Rx List. 2020. www.rxlist.com/toprol-xl-drug.htm#description

Varelius J. The value of autonomy in medical ethics. Med Health Care Philos. 2006;9(3):377–88.

Vivian JC. Criminalization of medication errors. US Pharm. 2009;34(11):66–8.

Vukadinovic D, Scholz SS, Messerli FH, et al. Peripheral edema and headache associated with amlodipine treatment: a meta-analysis of randomized, placebo-controlled trials. J Hypertens. 2019;37(10):2093–103.

Williams KG, Fellowss SE, Sanna LM. The role of the pharmacist in addressing the opioid crisis. Alb Gov't L Rev. 2017;11:174.

Chapter 8
Reproductive Ethics

Having laid the philosophical and historical groundwork for the ethics of pharmacy practice, we now turn to Section II of this text, where we consider specific issues and cases. We begin our discussion with reproductive ethics, which includes some of the most contentious topics of our modern era: contraception, induced abortion, and assisted reproductive technologies.

Contraception

Attempts to control a woman's fertility have an ancient pedigree going back centuries, with varying degrees of success. The modern era of pharmaceutical interventions for this purpose began in the early 1950s when Gregory Pincus and John Rock developed the first hormonal method of birth control. Popularly called "the pill," it used a combination of estrogen and progestin to inhibit ovulation while maintaining a woman's normal monthly menstrual flow. Marketing of the pill began in the U.S. in the 1960s, and at first, it was enormously successful. Yet the high estrogen content led to inconvenient side effects, such as headaches and weight gain. More serious was a much higher incidence of thromboembolic disease and stroke, as well as an increased risk for heart disease, especially in women who smoked. Modern formulations of oral contraceptives have much lower concentrations of hormones, with a more favorable side effect profile (Fig. 8.1) (Poston and Bouvier 2017).

For many American women, the practice of birth control is a practical necessity. It is vital for reproductive freedom and for the right of women to engage in sexual intimacy without the risk of pregnancy. Contraception promotes the equality of women and men and allows women to remain in the workplace. Furthermore, having children is expensive, and some couples do not have the financial resources for a large family. For those who place moral value on the unborn, having fewer children also reduces the risk of abortion.

Fig. 8.1 Model of a contraceptive pill, Europe, c. 1970. (Credit: Model of a contraceptive pill, Europe, c. 1970. Credit: Science Museum, London. Attribution 4.0 International (CC BY 4.0))

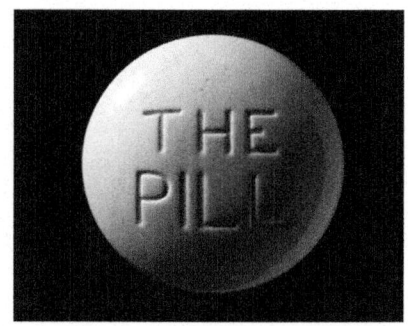

Religious objections to the morality of birth control are ancient, featured prominently in the history of the Roman Catholic Church. Many early church theologians opposed any interference with the "natural" conjugal act, including the thirteenth century church father Thomas Aquinas (Brown 2018). As we discussed in Chap. 2, the ethical theory of natural law is at the heart of these concerns. You may recall that natural law derives morality from a particular understanding of the nature of the universe and the created nature of human beings. On this view, hormonal birth control goes against the way God intended for procreation to take place. Pope Pius XI, in his 1930 encyclical *Casti Connubii,* provides a cogent twentieth century expression of this idea:

> Since, therefore, the conjugal act is destined primarily by nature for the begetting of children, those who in exercising it deliberately frustrate its natural power and purpose sin against nature and commit a deed which is shameful and intrinsically vicious. (Treacy 1939)

In addition to the practice of abstinence, the Pope gave approval for Catholic married couples to avoid pregnancy by engaging in intercourse only during the infertile times in a woman's monthly cycle (Devettere 2016). This so-called "rhythm method" is now referred to as natural family planning. Nonetheless, the Magisterium (teaching authority) of the Roman Catholic Church still teaches that hormonal contraception is immoral (Austriaco 2011). In addition to these theologically-based arguments, many conservative Catholics would make the utilitarian case that the ready availability of contraceptives promotes an immoral lifestyle and an increase in sexually-transmitted infections. It is worth noting that several conservative Protestant denominations also hold similar views. Liberal Protestant groups and various branches of Judaism and Islam are generally more accepting of contraceptive use (Pandia Health 2019).

In addition to these theological and religious discussions, there remains a final concern about hormonal contraceptives, that they might perhaps fail as *contraceptives,* and instead have a mechanism of action that actually destroys an early embryo and would therefore be *abortive.*

As we reviewed in Chap. 3, many conservative adherents to the major monotheistic religious traditions (Christianity, Judaism, and Islam) have a high view of human life as created in God's image. These groups would grant the unborn full moral status as persons from the time of fertilization/conception. Scientifically speaking, fertilization takes place high in the Fallopian tube, after which it takes up to 6 days for the developing human embryo to make its way down the tube for implantation into the fundus of the uterine wall (Tortora and Derrickson 2018). On the conservative view we are describing, anything that interferes with this trajectory is immoral since it destroys an early human person. Such a mechanism would only be operative after "breakthrough" ovulation, when the contraceptive mechanism fails, so it would have to be rare. In order to fully evaluate this moral concern, we must briefly review the primary mechanism of action of oral contraceptives.

The commonly-used modern pill is a combined oral contraceptive (COC) containing both estrogen and a progestin (Burkman et al. 2004; Gallo et al. 2005). Because of negative feedback, COCs inhibit the release of gonadotropin-releasing hormone (GnRH) from the hypothalamus. This action reduces the levels of luteinizing hormone (LH) and follicle-stimulating hormone (FSH). Without FSH, a dominant follicle does not develop in the ovaries. There is also no surge in LH at mid-cycle, so no signal for ovulation even if a follicle were present. COCs have the added effect of altering the cervical mucus, making it less hospitable for sperm penetration should ovulation occur (Han et al. 2017; Bucciero and Parda-Chlebowicz 2018; Sullivan 2006).

In addition to suppression of ovulation and interference with sperm, another possible COC mechanism of action is often mentioned by drug manufacturers. Consider this excerpt from the online description of the contraceptive product Yasmin®:

> COCs lower the risk of becoming pregnant primarily by suppressing ovulation. Other possible mechanisms may include cervical mucus changes that inhibit sperm penetration and endometrial changes that reduce the likelihood of implantation. (RxList 2020)

Overwhelming evidence shows that COCs work mostly by suppressing ovulation (Kovacs 2003). However, in the rare event that breakthrough ovulation occurs, changes in cervical mucus make it more difficult for sperm to travel through the reproductive tract. Both of these are actual *contraceptive* effects, preventing fertilization. The third possible effect would be an *interceptive* effect, where the endometrium may be less receptive to implantation of a newly-formed embryo (Sullivan 2006). For those who hold to a moral status for human embryos, this last mechanism leads to serious ethical concerns.

Herein lies the ethical dilemma for those who raise these objections. They often describe oral contraceptives as "abortifacient" (lit. "abortion-causing"), based on a rare but possible interceptive mechanism. It all depends on whether such a mechanism ever takes place, and this has been the subject of vigorous research and debate. Though some individuals informed by a religious viewpoint still believe that the pill interferes with implantation, research in the past 20 years or so has failed to conclusively detect any actual interceptive effect of simple combined oral contraceptives (Burkman et al. 2004; Rivera et al. 1999; Petitti 2003; Norwitz et al. 2001).

What about so-called emergency contraception, also known as the "morning-after pill?" *Emergency contraception* (EC) is intended as a necessary expedient to prevent pregnancy in women who have not been using other forms of birth control effectively or at all, or where barrier methods have failed (e.g., condom breakage).[1] Though there are several possible agents used for EC, the most common (and the main focus of this section) is the high-dose progestin levonorgestrel, marketed as Plan B One-Step®. Other marketed progestin-only forms include Next Choice One Dose®, My Way®, and Take Action® (Trussell and Raymond 2016). All of these forms of EC are available without a prescription (The Emergency Contraception 2020). If administered during the follicular phase of the menstrual cycle just before ovulation, levonorgestrel inhibits the release of LH from the anterior pituitary gland. This action may prevent or delay ovulation (Trussell and Raymond 2016; Baird 2009).

In several studies, levonorgestrel has demonstrated efficacy rates in a range of 52–94% (Arowojolu et al. 2002; von Hertzen and Godfrey 2009; Hamoda et al. 2004; Wai Ngai et al. 2005; Glasier et al. 2010; Segall-Gutierrez and Tilley 2010). Efficacy decreases with increasing amounts of time between coitus and medication usage (Trussell and Raymond 2016).[2] All of this assumes, of course, that EC methods were used only during the period of the woman's cycle just before or immediately after ovulation, or all other considerations are moot (i.e., the woman was not at risk to become pregnant in the first place). The efficacy rates are well established, but the mechanism of action has been poorly characterized. This has opened the door for considerable speculation about a possible post-fertilization effect, wherein the uterine endometrium may be rendered unreceptive for implantation.

Religious conservatives remain very concerned about such a possible interceptive mechanism of high-dose levonorgestrel. However, a few caveats are in order. The conservative animus against EC drugs began by automatically assuming that EC drugs were interceptive in their mechanism, at least some of the time. This automatic assumption has led to a curious "inversion" of the scientific method. In recent decades, unlike the typical course of discerning the mechanism of action of a drug, it seems to have been the burden of the medical and scientific community to prove how a drug *doesn't* act by proving beyond any doubt how it *does* act. Given that it is difficult to prove a negative, it seems more appropriate to rely on the large number of studies that have failed to demonstrate an interceptive mechanism of EC and leave it at that (Trussell and Raymond 2016; Emergency Contraception 2020; Peter 2017; Kahlenborn et al. 2002; Lalitkumar et al. 2007; Langston 2010; Meng et al. 2010). One of the authors of this text (DMS) has co-authored a review of this evidence (Lewis and Sullivan 2012). All of us are sufficiently motivated by the available data to believe that EC drugs possess no clinically relevant interceptive effect during the post-fertilization period.

More recently, a form of ulipristal acetate, marketed as ella®, has been approved by the FDA as an emergency contraceptive following unprotected sexual intercourse or barrier method failure (e.g., condom breakage). This EC agent acts as a potent modulator of progesterone receptors, leading to the reversible inhibition of

---

[1] Some parts of this section are modified from Lewis and Sullivan (2012); used with permission.
[2] *Plan B One-Step (Levonorgestrel) Package Insert.* 2009.

progestin. The mechanism of action varies, based on the time of administration in relation to a woman's menstrual cycle. When taken before ovulation, ulipristal may delay or inhibit ovulation. If taken after ovulation, ulipristal may cause decreased endometrial thickness, which might, therefore, cause an interceptive effect on embryo implantation if fertilization occurs (Whalen and Rose 2012). Such a mechanism of action would be of concern to pro-life religious conservatives.

However, one recent study showed that "ulipristal is not more effective at preventing pregnancy than chance alone if used after ovulation and does not increase miscarriage rates." The authors conclude that "an anti-implantation effect of [ulipristal] is highly unlikely at the dose used for EC." The authors comment that the continued use of an FDA-approved label indicating a possible interceptive effect may be counterproductive, since it might discourage women from using an otherwise highly effective form of birth control (Li et al. 2019).

In the previous chapter, we emphasized the need for pharmacists to practice the virtues articulated in the Code of Ethics for Pharmacists. Many sincere pharmacists may remain uncertain or unconvinced about the ethics of combined oral contraceptives or EC agents. If so, they may wish to act on their conscientious right of refusal to dispense such drugs, a topic we will take up in Chap. 11 of this text.

Abortion

Abortion is an ethical litmus test, dividing us along religious and political lines, leading to some of the most contentious and bitter debates of our modern era. Many social and religious conservatives have opposed all forms of abortion on absolutist grounds and claim that the human embryo/fetus should be inviolate from the moment of conception. Exceptions to this are rarely permitted and usually only apply when the life of the mother is at risk. Notably, these deeply-held beliefs cross traditional denominational boundaries, so that conservative Protestants have joined their conservative Roman Catholic counterparts in this pro-life stance. On the other hand, those favorable to the morality of abortion claim it is merely a woman's free choice, based on autonomy. The human embryo/fetus does not have moral status, or at least not enough to trump a woman's autonomy. How she handles an unwanted pregnancy is a private affair between her and her physician alone.

The question of abortion is a profoundly moral one because it involves another human life. As we discussed in Chap. 3, a *person* is "a member of the moral community" (Bagnoli 2007). The unborn embryo or fetus is undeniably genetically human, yet does it fit this description? This question may never be objectively resolved and may well be fruitless, given the bias and preconceptions that all sides bring to it. Devetterre summarizes the two poles of the debate in this way:

> Prenatal life ... raises questions about two important human goods: (1) the woman's personal choices and responsibility for her life and (2) the important reality of a distinctly new human life. It is precisely this dual nature of prenatal life, of course, that creates the major moral dilemmas. (Devettere 2016)

We begin our ethical analysis with a survey of major legal precedents, the first a contraception case that would later influence abortion law. In the early 1960s, Estelle Griswold, the Director of Planned Parenthood of Connecticut, defied a long-standing state law banning contraception counseling. She appealed her criminal conviction for this offense, asking if the U.S. Constitution protected the right of marital privacy from such restrictions. In 1965, the case was decided by the U.S. Supreme Court in *Griswold v. Connecticut*. Associate Justice William O. Douglas wrote for the majority in the 7-2 decision invalidating the Connecticut law, using an interesting and controversial argument to make his case. Although the Constitution does not explicitly grant a right to privacy, he held that "specific guarantees in the Bill of Rights have penumbras, formed by emanations from those guarantees that help give them life and substance" (U.S. Supreme Court 1965). Though later ridiculed by conservatives, Griswold formed the basis for a constitutional right of privacy that was persuasive enough to become a strong precedent for later decisions (Oyez 1965), as we will see (Fig. 8.2).

In 1969, a young Texas woman, later identified only as "Jane Roe," found herself in her third pregnancy, having given up her first two babies. Claiming she was raped, she wanted an abortion, which was illegal in Texas except to save a woman's life. Though her request was denied and the resulting baby was given up for adoption, the case began a long journey through the courts. The claim of rape was later found to be false (Devettere 2016).

In 1973, the U.S. Supreme Court rendered its decision in *Roe v. Wade*, again by a 7-2 decision. Associate Justice Harry Blackmun delivered the opinion for the

Fig. 8.2 The Supreme Court of the United States, Washington, DC. (Photo by Mr. Kjetil Ree – own work, CC BY-SA 3.0)

majority. He argued that the right of privacy previously established in *Griswold* covered a woman's right to an abortion. His lengthy analysis employed the trimester system, dividing pregnancy into three periods of 3 months each. For the first 12 weeks, a woman had the complete and total decisional authority to decide whether or not to terminate her pregnancy. There were different levels of possible state involvement in the second and third trimesters. Roe became a massive precedent that changed the laws in 46 states (Oyez 1973).

Recognizing that it would have far-reaching implications, Justice Blackmun's *Roe* opinion sought to address the controversies relating to human personhood in the womb. He wrote: "If this suggestion of personhood is established, the appellant's case, of course, collapses, for the fetus' right to life is then guaranteed specifically by the [Fourteenth] Amendment" (U.S. Supreme Court 1973). The Fourteenth Amendment declares that no "state [may] deprive any person of life, liberty, or property, without due process of law" (America's Founding Documents 2020). Justice Blackmun went on to acknowledge the importance of the personhood question but then declined to rule on that basis: "We need not resolve the difficult question of when life begins. When those trained in the respective disciplines of medicine, philosophy, and theology are unable to arrive at any consensus, the judiciary, at this point in the development of man's knowledge, is not in a position to speculate as to the answer" (U.S. Supreme Court 1973). The Court, therefore, clearly decided to make autonomy, defined as a woman's right to privacy, the central issue. Thus the right to abortion became firmly established as a judicial precedent.

In 1992, the Court reached a tightly-contested 5-4 decision in *Planned Parenthood of Southeastern Pennsylvania v. Casey*. In an unusual joint opinion written by justices O'Connor, Kennedy, and Souter, the Court retained the essential elements of *Roe*, while discarding the trimester system, noting that medical advances had made fetal viability possible at a much earlier stage of pregnancy. But the Court threw out a spousal notification requirement, declaring that it imposed an "undue burden," defined as a "substantial obstacle in the path of a woman seeking an abortion before the fetus attains viability" (Oyez 1992). This idea of an undue burden has become a major determinant of the constitutionality of state or federal restrictions on abortion ever since *Roe* and *Casey* (Shimabukuro 2019).

Yet even the undue burden standard has its limits. In *Gonzales v. Carhart* (2007), the U.S. Supreme Court upheld, by a narrow 5-4 decision, a federal law banning a specific type of late-term abortion called "partial-birth abortion." More technically known as *intact dilation and extraction*, the procedure employs a feet-first delivery, followed by puncture decompression of the skull to permit easier removal of the now-dead fetus. Writing for the majority, Associate Justice Anthony Kennedy argued that the ban on the procedure was very specific, therefore not unconstitutionally vague, and that it did not impose an undue burden on the right to an abortion. Against the argument that an exception was needed for abortions protecting the life of the mother, the majority was convinced by expert testimony that intact dilation and extraction was never actually medically necessary (Oyez 2007).

At this point in our discussion, it is worth examining the methods used to accomplish an elective abortion. In the first 16 weeks of pregnancy, the most common method is suction aspiration, which uses a powerful suction tube with a sharp cutting edge inserted into the uterus to remove the fetus and placenta (American Pregnancy Association 2021). After 16 weeks of gestation, abortion is much less common. By that time, the procedure must be modified to allow for the size of the developing fetus. In the procedure called ***dilation and evacuation*** (not to be confused with partial-birth abortion), the cervix is dilated under local anesthesia, and forceps are used to remove fetal body parts, with the endometrial lining removed with a curette. Possible complications from both procedures may include nausea, bleeding, and (less likely) cervical damage, uterine puncture, or infection (American Pregnancy Association 2021).

It is also possible to perform an early abortion without surgery. A medical abortion can be performed at up to 10 weeks of gestation, involving the use of two medications. The first is mifepristone, a progesterone-receptor antagonist. Progesterone's action at endometrial receptors stimulates blood flow in the arterioles that supply blood to the decidua and ultimately to the placenta. Mifepristone blocks the action of progesterone, which interferes with the decidual blood supply, causing fetal loss. A gynecologist typically administers mifepristone to the patient in the office or clinic (Justine 2010). The second medication is misoprostol, a prostaglandin that induces uterine contractions, often dispensed by a community pharmacist. The patient takes misoprostol 24–48 h after the first medication, which then completes the abortion procedure. Possible complications include bleeding, retained tissues, or infection, necessitating antibiotics or a follow-up surgical procedure (Planned Parenthood 2020).

Abortion Ethics

Opponents of elective abortion use a variety of arguments to make their case. Pro-life advocates are convinced that the unborn entity is a person. To express this idea in religious language, the embryo/fetus is made in God's image and possesses full moral status that warrants protection. The absolutist view holds that all abortions are therefore immoral and that the unborn entity has an absolute right to life. Taking this position to an extreme, however, reveals some potential problems.

Consider the example of an ectopic pregnancy, a relatively uncommon condition where an early embryo prematurely implants into the wall of the Fallopian tube, rather than into the endometrium of the uterus itself. There is no possibility that the embryo/fetus will survive, and the situation is likely to become dangerous to the mother. At 8–9 weeks of gestation, the gestational sac will enlarge to the point of tubal rupture, which also tears the ovarian artery. The resulting massive hemorrhage in the pelvic region places the life of the mother at serious risk. Therefore, a ruptured ectopic pregnancy is a surgical emergency.

It is clearly in the best interests of the patient to remove such a pregnancy before tubal rupture, usually performed as a laparoscopic procedure. However, based on deontological theories such as Kantian ethics or divine-command, some would view such an elective procedure as morally problematic, since it destroys a living embryonic human person. However, from a utilitarian perspective, operating to remove such a pregnancy is appropriate, as it protects the life of the mother, and there is no way to save the embryo/fetus. Though often impermissible to perform such a procedure in a Catholic hospital, it is nonetheless the preferred outcome, according to the vast majority of clinical observers.

Let us now consider the other side of the moral equation. Many observers of the pro-choice persuasion hold that the unborn entity has no moral status and that the choice of whether or not to continue a pregnancy is strictly up to the mother and her doctors. An extreme form of this libertarian view would therefore condone abortion at any time during pregnancy, even in the third trimester immediately before birth. However, there is a serious potential problem with this view as well.

Utilitarian philosophers Peter Singer and Helga Kuhse took this idea to its logical conclusion in their well-known 1985 book: *Should the Baby Live?* (Kuhse and Singer 1985). They argued for the morality of putting disabled infants to death. As discussed in more detail in Chap. 3, they advocated for the moral equivalency of the pre-born fetus and the newborn: since (on their view) if neither is self-aware, then neither is a person. Therefore, this approach holds that infanticide is substantially equivalent to abortion and is morally acceptable. Nonetheless, very few observers across the social and political spectrum would agree with this extreme position.

We cannot, in this summary, pursue the various ethical positions on abortion in more detail. However, given these extremes on both sides of the debate, most clinicians are somewhere in the middle: a large gray zone of uneasy assumptions and an intuitive sense that the entity in the womb has greater and greater moral status as the pregnancy progresses.

Assisted Reproductive Technologies

Infertility affects about 15% of all couples in the United States, defined as a failure to achieve a viable pregnancy after 1 year of trying (Thoma et al. 2013). Medical interventions to help infertile couples are centuries old, with the first actual attempts at artificial insemination in the early nineteenth century. *Artificial insemination*, now usually referred to as *intrauterine insemination* (IUI), has the goal of increasing gamete density at the site of fertilization. The first routinely successful techniques for IUI date back to the 1940s. The method has been commonly employed in cases of male-factor infertility due to erectile dysfunction or oligospermia (low sperm count) (Ombelet and Van Robays 2015).

When the problem is more on the female partner's side and surgical correction is not feasible, the couple may make use of *in vitro* fertilization (IVF). This newer method became a reality when Louise Joy Brown, now universally famous as the world's first "test-tube baby," was born. Her parents, Leslie and John Brown, were unable to naturally conceive for 9 years, due to blockages in the mother's Fallopian tubes. In 1978, Patrick Steptoe and Robert Edwards laparoscopically retrieved a single, non-stimulated oocyte from Leslie. The doctors used the husband's sperm to fertilize the oocyte in the lab, and the resulting embryo was transferred into Leslie's uterus several days later. On July 25th, Louise Brown's birth by cesarean section ushered in a new era of assisted reproductive technologies. In 2010, Robert Edwards received a Nobel Prize for his contributions to this medical breakthrough (Ombelet and Van Robays 2015; Kamel 2013).

The current technique for IVF involves using hormones functionally similar to FSH to hyperstimulate the woman's ovaries. If successful, multiple oocytes are retrieved via transcutaneous needle aspiration with ultrasound visualization. These oocytes are then combined in a Petri dish with the male partner's ejaculated sperm or with donated sperm, so that fertilization may take place. Resulting embryos that reach the blastocyst (5 day) stage can then be transferred directly into the woman's uterus to implant and possibly achieve a viable pregnancy. A variation of this technique, intracytoplasmic sperm injection (ICSI), is used in cases of male oligospermia. This method uses a single sperm cell injected directly into the cytoplasm of an oocyte to achieve fertilization (Fig. 8.3) (Ness and Montvilo 2018).

IUI, IVF, and ICSI are collectively known as *assisted reproductive technologies* (ART). There are a large number of ethical issues, briefly summarized here. To begin with, the very idea of extra-uterine conception runs counter to the perspective of natural law. The advent of ART has been so disturbing to the Magisterium of the Roman Catholic Church that a new directive was published by the Vatican in 2008, entitled "Instruction *Dignitas Personae* on Certain Bioethical Questions" (Levada and Ladaria 2008). One Catholic scholar commented on the directive this way: "assisted reproductive technology involves infertility doctors *reproducing* human life, not spouses *procreating* it" (Latkovic 2010). On this view, ART is immoral because it is unnatural, going against the way God created human beings. Human beings are merely reproductive products, rather than procreated persons arising out

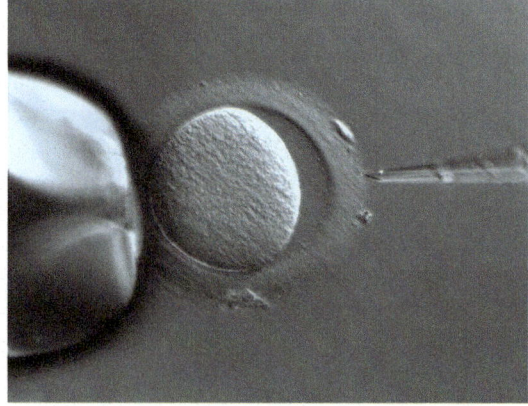

Fig. 8.3 Human oocyte with zona pellucida, *in vitro* fertilization. (Illustration from ZEISS Microscopy from Germany – Oocyte with Zona pellucida. CC BY-SA 2.0)

of the conjugal relationship of sexual intimacy. The overriding concern is for the commodification and depersonalization of the resulting progeny.

> **Assisted Reproductive Technologies (ART)**
> - intrauterine insemination (IUI)
> - *in vitro* fertilization (IVF)
> - intracytoplasmic sperm injection (ICSI)

There are other ethical issues as well that are not strictly religious. Since IVF produces a large number of human embryos, what happens to the unimplanted excess ones left over? Modern technology can cryopreserve them for later use. But what is the status of such "frozen" embryos? Even if one does not hold them to be full moral persons, most clinicians believe that they are unique in some way, more than mere tissue samples. What might happen to them? There are at least five possible outcomes for unused frozen embryos.[3] First, they may be implanted at a later time in the wombs of those who provided gametes for their creation. Some who hold a high view of embryos often recommend that *only* those embryos that will eventually be implanted be created in the first place (Elkins 2000), preventing the problem. The second alternative would be to release frozen embryos for implantation into another womb. Such embryo donation, sometimes even referred to as "embryo adoption" can provide childless couples the joy of having a baby, while at the same time acting on behalf of another (Mitchell 2000). Embryo donation is readily available to those who wish to pursue this idea. The success rate for achieving pregnancy and carrying a child to term is similar to that of routine IVF (Check et al. 2004). From a legal perspective, embryos are property, whose disposition is the sole prerogative of the "owners." Yet the law has been ambiguous, often claiming that embryos are "irreplaceable" and "unique." Current law is conflicting and arbitrary regarding IVF and the embryos derived from it (Glenn 2002).

So, setting aside the procreative options for the moment, the third option would be to leave frozen embryos in their current state of physical, moral, and legal limbo, with the parents/owners unwilling or unable to decide their fate. Of course, this sort of non-decision may eventually lead to the destruction of the embryos, but the limit of viable long-term storage of frozen embryos is not yet known. The fourth choice is to request the elimination of the frozen embryos. By far, this is the most common outcome, by an explicit decision of the parents/owners. In a survey of 1246 couples who had decided to no longer try for a pregnancy themselves, the vast majority (89.5%) requested the destruction of their frozen embryos, even though other options were available (Kovacs et al. 2003). Another survey of over 3800 couples revealed that 9.1% would be willing to donate their embryos to another couple for adoption (Moutel et al. 2002), while a separate study of 509 couples revealed that

---

[3] Portions of this section are modified from Sullivan and Costerisan (2008). Used with permission.

about 10% would consider donating their embryos for stem-cell research, the fifth and final option (McMahon et al. 2003).

> **Possible Outcomes for Unused Frozen Embryos**
> 1. Implantation into womb of parent/owner
> 2. Implantation into another womb (embryo donation)
> 3. Leave frozen embryos in their current state (unknown limit for viable long-term storage)
> 4. Embryo destruction (most common outcome)
> 5. Embryo donation for stem-cell research

What if the created embryos have genetic issues? Many infertility clinics are increasingly utilizing preimplantation genetic diagnosis (PGD). The primary purpose of PGD is to identify and select genetically healthy embryos for implantation in parents who may be carriers of specific genetic abnormalities. The technique involves biopsy of preimplantation embryos at the 8-cell stage, followed by an analysis of the removed cells. Genetically healthy embryos can then be implanted while discarding those with genetic disorders (Sueoka 2016).

Indications for PGD screening include a variety of monogenic (single gene mutation) diseases, such as cystic fibrosis, sickle cell anemia, beta-thalassemia, and spinal muscular atrophy, all of which are relatively common autosomal recessive disorders. Certain less common autosomal dominant diseases are also targets for analysis, including myotonic dystrophy, Huntington's disease, and Marfan's syndrome. The list also includes X-linked diseases, such as Duchenne muscular dystrophy and hypoxanthine-guanine phosphoribosyltransferase (HPRT) deficiency. Finally, PGD is often used to detect aneuploidy conditions, where an embryo may have too many or too few chromosomes. The most common example is Down syndrome (Trisomy 21) (Sermon et al. 2004).

Down syndrome provides a good starting point to discuss the ethics of PGD. In Trisomy 21, the extra genetic material leads to a small stature and other physical characteristics and may lead to cognitive disabilities as well, usually mild. The National Down Syndrome Society has said this:

> People with Down syndrome have an increased risk for certain medical conditions such as congenital heart defects, respiratory and hearing problems, Alzheimer's disease, childhood leukemia, and thyroid conditions. Many of these conditions are now treatable, so most people with Down syndrome lead healthy lives...
>
> All people with Down syndrome experience cognitive delays, but the effect is usually mild to moderate and is not indicative of the many strengths and talents that each individual possesses. (National Down Syndrome Society 2020)

From a conservative perspective, all human embryos have moral status, and objections to discarding affected embryos mirror concerns about abortion generally. In addition, many would claim that destroying embryos for Down syndrome is genetic discrimination and a modern form of eugenics, the flawed and discredited pseudo-science we discussed in Chap. 5.

The case of Molly Nash provides a controversial additional ethics dilemma. Six-year-old Molly was suffering from Fanconi's anemia (FA), a genetically inherited immune deficiency that leads to an early death, usually by age seven. She was very symptomatic and had received multiple blood transfusions. In desperation, her parents turned to IVF and PGD to create a "savior sibling," whose umbilical cord blood was not only negative for the FA abnormality but also a tissue match for Molly. After multiple cycles where FA embryos were discarded, and other embryos were unsuccessfully implanted, one embryo finally resulted in a pregnancy. Shortly after Adam Nash was born, doctors transplanted his umbilical cord cells into Molly, which repopulated her bone marrow and (assuming the durability of the grafted cells) permanently cured her anemia (Dobson 2000; Haude et al. 2017).

A potential ethical objection in this scenario is the commodification of human life. As we reviewed in Chap. 2, the second categorical imperative of Kantian ethics states that persons should be ends in and of themselves, never the means to another person's ends. At least in part, Adam was a 'means' rather than an 'end.' While the parents involved strongly claimed that they loved and desired Adam in his own right, this concern seems well warranted (Sermon et al. 2004; Haude et al. 2017).

The case of Molly Nash brings to mind another ethical concern related to various forms of ART coupled with PGD. So far, we have discussed possible *negative* eugenics implications of this technology that would eliminate embryos with negative traits, but there may be *positive* eugenics ideas here as well. You may recall that Chap. 5 described the "fitter family" and "better baby" contests from the 1920s, sponsored by the American Eugenics Society. Such examples of positive eugenics relied on appropriate marriage partners to achieve certain desirable traits, such as high intelligence or athletic abilities (Pernick 2002).

With today's advanced reproductive technologies in view, rather than eliminating unwanted genetic traits leading to disease, PGD could be employed to select out highly-desirable embryos, or even merely to choose a specific gender. Of course, a "positive" eugenics plan could have negative implications, for this would lead to the discarding of unwanted human embryos. The idea of using genetic screening to create such "designer babies" has met with a variety of ethical reactions, ranging from skepticism to outright condemnation (Verlinsky 2005; Handyside 2010; Cook 2017; Segers et al. 2019).

A final ethics issue revolving around IVF and other forms of ART is the high cost and significant uncertainties regarding insurance coverage. A recent study reached these conclusions:

> [C]urrent insurer and governmental policies pose a series of ethical and logistical challenges for patients and providers, including multiple uncertainties concerning costs and coverage for ART, e.g., regarding the cause, length, prices, and outcomes of infertility treatments, and the likelihood that insurers will cover any particular procedure. Patients must hence often pay first, and later attempt to get reimbursed. But insurers frequently will not commit to coverage in advance, and instead make decisions only afterward, on a case-by-case basis, which can take months. Patients and providers thus cannot predict in advance

how best to allocate limited coverage amounts. Given these uncertainties, many patients adopt several strategies, and providers may seek to advocate for patients with insurers, but are usually unsuccessful. Patients, who continue to try to have a child, face quandaries of how much to keep spending – how much a child is worth – and must make complex risk/benefit decisions. (Klitzman 2017)

Infertile Americans without adequate insurance coverage may not be able to afford ART, since the charge for even one cycle of IVF may be more than $30,000. If conceiving a child by such means is only for the rich, this raises serious concerns about distributive justice. In desperation, some may even seek such treatments in other countries such as Thailand or India, a trend often referred to as "reproductive tourism," with concerns about safety, success rates, and the exploitation of local economies (Inhorn and Patrizio 2009).

ART and Surrogacy

We cannot leave our in-depth discussion of reproductive ethics without considering the question of surrogate parenting. Consider the following case history, taken from a BBC News report:

Bobby and Nikki Bains from Ilford, England, were both in their 40s. At the time of this news story, their three-month-old daughter Daisy was the product of Bobby's sperm and a donor egg, brought together in a Petri dish through IVF. The resulting embryo was then implanted into the womb of a woman the Bains had never met. All this took place in the Rotunda Clinic in Mumbai, India. After the child was born, she was turned over to the British couple, and the surrogate mother was paid a fee for her services. (Taneja 2008)

Surrogate motherhood is ethically controversial. It is legal in the U.K., though heavily regulated, which is why the Bains may have chosen to go elsewhere. In India, often called the "surrogacy hub" of the world, such arrangements are widespread and growing in popularity, with no shortage of women willing to act as surrogates. The Indian government has recently discussed banning the practice due to concerns over the exploitation of vulnerable women, especially given their high maternal mortality rate (*BBC News* 2016).

Here in the United States, the surrogacy issue first hit the headlines in the 1980s, with the famous "Baby M" case. William Stern signed a contract with a woman willing to become a surrogate mother. His wife was infertile, but they both desperately wanted a child. Their agreement with Mary Beth Whitehead promised to pay her $10,000 plus medical expenses to gestate a child through artificial insemination (more recently, such arrangements are almost always carried out through IVF).

In March of 1986, Mary Beth Whitehead gave birth to a little girl, the biological product of her own egg, with sperm from Mr. Stern. In defiance of her contract, she named the child Sara and refused to give her up. The Sterns, who had called the child Melissa, sought a court order to get her back, but rather than surrender the child, the Whiteheads fled to Florida. Eventually, they were caught, and the baby was returned to the Sterns in New Jersey. A court awarded custody of the baby to the

Sterns because they were thought to be better parents than the Whiteheads, though they received limited visitation rights (Feldman 2018).

One reason that a case like this can be confusing is that surrogacy allows for different kinds of parents. For example, a woman can be the *biological mother* of a child if her egg is fertilized. But with modern reproductive technologies, the *gestational mother* may well be someone else, that is, the one who actually gives birth to the child. And the *nurturing mother* might be someone else entirely, the one who will actually be the parent and raise the child.

Advocates for this kind of arrangement point to our American tradition of reproductive liberty. As we have seen earlier, several Supreme Court precedents establish a constitutional right to make one's own decisions about childbearing. However, such precedents refer most directly to the right to birth control and the right to an abortion. It is unclear whether such decisions address the issue of surrogacy, as the current plethora of conflicting states laws will attest:

> Is surrogacy in the United States Legal? Surrogacy in the United States remains unregulated at the federal level, with each individual state having its own laws (or not). The individual state laws vary widely even between states that are considered "surrogacy friendly." (Houghton 2020)

Setting aside legal concerns, what about the ethics of all this? From a religious standpoint, conservative Judeo-Christian and Muslim traditions have voiced grave concerns about the morality of surrogate motherhood. For one reason, the book of Genesis describes the marriage bond as "one flesh."[4] Now, this may mean a variety of things, but this text has caused some religious ethicists to rule out bringing a separate "third party" (the surrogate mother) into a marriage relationship. By the way, this same analysis influences the religious skepticism of "third party" gametes (anonymously donated sperm or eggs), in reproductive technologies generally.

Even if we set aside religious concerns, surrogacy arrangements may seem a lot like baby-selling. The cash incentives are lucrative and attractive for poor mothers, so this is more than just a fee for a service. This problem is particularly apparent, considering that surrogacy contracts may promise a small fraction of the promised total fee for a pregnancy ending in miscarriage – they only award the full price for a live, full-term baby. This idea of babies as more of a product than a person is another example of the commodification of human beings, with a resultant loss of dignity.

The potential ethical pitfalls are worrisome. What if a surrogate mother undergoes amniocentesis at 14 weeks of pregnancy, and the test determines that she is carrying a child with Down syndrome? Will the adoptive parents consider this a less than perfect baby and try to force her to have an abortion? What if the mother has bonded with the child during the pregnancy, as happened in the Baby M case, and she wants to keep the baby? Should the child be ripped from her arms merely because a legal contract says so?

---

[4] Genesis 2:24.

The preceding analysis derives from the idea of *commercial surrogacy*, where the surrogate mother is paid for her services and gives up her rights. Another form of surrogacy contract, called *altruistic surrogacy*, is a volunteer arrangement that may mitigate some of the ethical concerns. Imagine that the younger sister or cousin of an infertile woman lovingly offers to gestate her baby, using IVF and gametes from the infertile woman and her partner. The only payment involves covering the surrogate's medical expenses, who, after the delivery, becomes an especially close "aunt" to the baby. Though a few potential ethical problems remain, many would herald this idea as more acceptable (Baker 1996; Rao 2012; Bhattacharjee 2016).

Types of Surrogacy

- Commercial surrogacy: legal contract for pay, surrogate mother gives up rights.
- Altruistic surrogacy: volunteer arrangement, usually with a family member.

When all is said and done, the authors of this text remain skeptical of the advisability of surrogacy arrangements. For infertile couples who cannot themselves utilize ART, adoption may be a better choice.

Conclusion

This chapter has made a detailed examination of the highly divisive ethical issues involved with human reproduction. We began with hormonal contraception, considering both standard and emergency forms of pregnancy prevention, viewing these through the lenses of natural law, pro-life concerns, and women's reproductive choice. We then considered the huge debates over abortion, beginning with the legal precedents. We then looked at the ethical extremes on both poles, with most clinicians occupying an uneasy middle ground. Finally, we examined the new arena of reproductive technologies. We considered the ethical concerns surrounding IVF and other forms of interventions, such as preimplantation genetic diagnosis and surrogacy arrangements, to help infertile couples to conceive.

The next two chapters will consider ethics at the end of life.

Key Terms

- Combined oral contraceptive
- Emergency contraception
- Contraceptive v. interceptive effects
- Natural family planning
- Major U.S. Supreme Court Decisions:

- *Griswold v. Connecticut* (1965)
- *Roe v. Wade* (1973)
- *Planned Parenthood of Southeastern Pennsylvania v. Casey* (1992)
- *Gonzales v. Carhart* (2007)

- Suction aspiration (vacuum curettage)
- Dilation and evacuation
- Intact dilation and extraction
- Medical abortion
- Ectopic pregnancy
- Oligospermia
- Assisted reproductive technologies
  - intrauterine insemination (IUI)
  - *in vitro* fertilization (IVF)
  - intracytoplasmic sperm injection (ICSI)
- Preimplantation genetic diagnosis
- Savior sibling
- Reproductive tourism
- Commercial and altruistic surrogacy

Questions for Review and Discussion

1. What bearing does the idea of personhood (cp. Chap. 3) have on one's view of reproductive ethics?
2. What are the major arguments on both sides of the debate on the ethics of abortion?
3. When discussing the ethics of contraceptive agents, should the mechanism of action matter? Why or why not?
4. Does the use of modern assisted reproductive technology, specifically including preimplantation genetic diagnosis, qualify as a modern form of eugenics? Under what circumstances would this term apply?
5. In your opinion, what forms of reproductive surrogacy are ethical, if any? What ethical theories underlie your answer?

For Further Reading

Judith Jarvis Thomson. A Defense of Abortion. *Philosophy and Public Affairs*. 1971;1(1):47–66 (Thomson 1971). This classic essay, one of the most quoted philosophical works in the English language, is an attempt to justify abortion ethically, even if one assumes that an unborn fetus is a person. The essay features the famous "unconscious violinist" argument discussed in many ethics classes. Easily found on the Internet, this illustration is well worth reading and debating.

George W. Bush, Presidential Speech on Stem Cell Research, from Crawford, Texas (August 9, 2001) (Bush 2001).

Barack Obama, Presidential Speech on Stem Cell Research, from the White House (March 9, 2009) (Obama 2009).

It is highly instructive to read, side by side, these two key policy addresses by recent U.S. presidents. The first gives President Bush's reasons for forbidding NIH funding of embryo-destructive research. The second offers President Obama's reasons for overturning that ban. These speeches are easily found on the Internet – try to discern the key ethical theories that underlie each approach.

Jodi Picoult, *My Sister's Keeper: A Novel*. Washington Square Press, 2004. This book is an accessible fictional account of a family that uses IVF to create a "savior sibling" similar to Adam Nash, with many plot twists. The book was eventually made into a movie.

References

American Pregnancy Association. Surgical abortion. American Pregnancy Association Web site. https://americanpregnancy.org/unplanned-pregnancy/abortion-75934/. Published 2021. Accessed 15 Feb 2021.
America's Founding Documents. Fourteenth Amendment. America's Founding Documents Web site. www.archives.gov/founding-docs/amendments-11-27. Published 2020. Accessed 15 Feb 2021.
Arowojolu AO, Okewole IA, Adekunle AO. Comparative evaluation of the effectiveness and safety of two regimens of levonorgestrel for emergency contraception in Nigerians. Contraception. 2002;66(4):269–73.
Austriaco NPG. Biomedicine and beatitude: an introduction to Catholic bioethics. Washington, DC: Catholic University of America Press; 2011.
Bagnoli C. Respect and membership in the moral community. Ethical Theory Moral Pract. 2007;10(2):113–28.
Baird DT. Emergency contraception: how does it work? Reprod Biomed Online. 2009;18(Suppl 1):32–6.
Baker BM. A case for permitting altruistic surrogacy. Hypatia. 1996;11(2):34–48.
BBC News. India unveils plans to ban surrogacy. BBC News. 2016. www.bbc.com/news/world-asia-india-37182197. Published 25 Aug 2016.
Bhattacharjee D. Commercial surrogacy in India. Econ Polit Wkly. 2016;51(14):27.
Brown CM. Thomas Aquinas. Internet Encyclopedia of Philosophy. 2018. www.iep.utm.edu/aquinas/
Bucciero M, Parda-Chlebowicz M. Contraception: overview. In: Ambulatory gynecology. New York: Springer; 2018. p. 33–57.
Burkman R, Schlesselman JJ, Zieman M. Safety concerns and health benefits associated with oral contraception. Am J Obstet Gynecol. 2004;190(4 Suppl):S5–22.
Bush GW. Bush announces position on stem cell research. White House Arch. 2001; https://georgewbush-whitehouse.archives.gov/news/releases/2001/08/20010809-2.html
Check JH, Wilson C, Krotec JW, Choe JK, Nazari A. The feasibility of embryo donation. Fertil Steril. 2004;81(2):452–3.
Cook M. Ethics for an edited embryo. Australas Sci. 2017;38(5):49.
Devettere RJ. Practical decision making in health care ethics: cases, concepts, and the virtue of prudence. 4th ed. Washington, DC: Georgetown University Press; 2016.
Dobson R. "Designer baby" cures sister. BMJ (Clinical research ed). 2000;321(7268):1040.
Elkins T. A medical educator's perspective. In: Kilner JF, Cunningham PC, Hager WD, editors. The reproduction revolution. Grand Rapids: William B. Eerdmans; 2000.
Feldman EA. BabyM turns 30: the law and policy of surrogate motherhood. Am J Law Med. 2018;44(1):7–22.

References

Gallo MF, et al. 20 mcg versus >20 mcg estrogen combined oral contraceptives for contraception. Cochrane Database Syst Rev. 2005;2:CD003989.

Glasier AF, Cameron ST, Fine PM, et al. Ulipristal acetate versus levonorgestrel for emergency contraception: a randomised non-inferiority trial and meta-analysis. Lancet. 2010;375(9714):555–62.

Glenn LM. Loss of frozen embryos. AMA J Ethics. 2002;4(12)

Hamoda H, Ashok PW, Stalder C, Flett GMM, Kennedy E, Templeton A. A randomized trial of mifepristone (10 mg) and levonorgestrel for emergency contraception. Obstet Gynecol. 2004;104(6):1307–13.

Han L, Taub R, Jensen JT. Cervical mucus and contraception: what we know and what we don't. Contraception. 2017;96(5):310–21.

Handyside AH. Twenty years of designer babies. Reprod BioMed Online (Reproductive Healthcare Limited). 2010;20:S1.

Haude K, McCarthy Veach P, LeRoy B, Zierhut H. Factors influencing the decision-making process and long-term interpersonal outcomes for parents who undergo preimplantation genetic diagnosis for Fanconi Anemia: a qualitative investigation. J Genet Couns. 2017;26(3):640–55.

Houghton W. Surrogacy in the United States. The Sensible Surrogacy Guide Web site. www.sensiblesurrogacy.com/surrogacy-in-the-united-states/. Published 2020. Accessed 13 Nov 2020.

Inhorn MC, Patrizio P. Rethinking reproductive "tourism" as reproductive "exile". Fertil Steril. 2009;92(3):904–6.

Justine W. Medication abortion using mifepristone and misoprostol. New York: Springer New York; 2010:347.

Kahlenborn C, Stanford JB, Larimore WL. Postfertilization effect of hormonal emergency contraception. Ann Pharmacother. 2002;36(3):465–70.

Kamel RM. Assisted reproductive technology after the birth of Louise Brown. J Reprod Infertil. 2013;14(3):96.

Klitzman R. How much is a child worth? Providers' and patients' views and responses concerning ethical and policy challenges in paying for ART. PLoS One. 2017;12(2):e0171939.

Kovacs GT. Pharmacology of progestogens used in oral contraceptives: AN historical review to contemporary prescribing. Aust N Z J Obstet Gynaecol. 2003;43:4–9.

Kovacs GT, Breheny SA, Dear MJ. Embryo donation at an Australian university in-vitro fertilisation clinic: issues and outcomes. Med J Aust. 2003;178(3):127–9.

Kuhse H, Singer P. Should the baby live? The problem of handicapped infants. Oxford: Oxford University Press; 1985.

Lalitkumar PGL, Lalitkumar S, Meng CX, et al. Mifepristone, but not levonorgestrel, inhibits human blastocyst attachment to an in vitro endometrial three-dimensional cell culture model. Hum Reprod. 2007;22(11):3031–7.

Langston A. Emergency contraception: update and review. Semin Reprod Med. 2010;28(02):095–102.

Latkovic MS. The dignity of the person an overview and commentary on dignitas personae. Natl Catholic Bioeth Q. 2010;10(2):283–305.

Levada WC, Ladaria L. Instruction dignitas personae on certain bioethical questions. Vatican City: Congregation for the Doctrine of the Faith; 2008.

Lewis JD, Sullivan DM. The abortifacient potential of emergency contraceptives. Ethics Med. 2012;28:113–20.

Li HWR, Resche-Rigon M, Bagchi IC, Gemzell-Danielsson K, Glasier A. Does ulipristal acetate emergency contraception (ella®) interfere with implantation? Contraception. 2019;100(5):386–90.

McMahon CA, Gibson FL, Leslie GI, Saunders DM, Porter KA, Tennant CC. Embryo donation for medical research: attitudes and concerns of potential donors. Hum Reprod (Oxford England). 2003;18(4):871–7.

Meng C-X, Marions L, Byström B, Gemzell-Danielsson K. Effects of oral and vaginal administration of levonorgestrel emergency contraception on markers of endometrial receptivity. Hum Reprod. 2010;25(4):874–83.

Mitchell CB. NIH, Stem Cells, and Moral Guilt. Center for Bioethics and Human Dignity. 08/23/2000. https://pdfs.semanticscholar.org/1da5/483050bbc329414aa397c711eba4915fc622.pdf

Moutel G, Gregg E, Meningaud JP, Herve C. Developments in the storage of embryos in France and the limitations of the laws of bioethics. Analysis of procedures in 17 storage centres and the destiny of stored embryos. Med Law. 2002;21(3):587–604.

National Down Syndrome Society. Down syndrome facts. National Down Syndrome Society Web site. www.ndss.org/Down-Syndrome/Down-Syndrome-Facts/. Published 2020. Accessed 20 Dec 2020.

Ness BP, Montvilo RKP. Assisted reproductive technologies. Salem Press; 2018.

Norwitz ER, Schust DJ, Fisher SJ. Implantation and the survival of early pregnancy. N Engl J Med. 2001;345(19):1400–8.

Obama B. Presidential remarks on stem cell research. White House Arch. 2009; https://obamawhitehouse.archives.gov/the-press-office/remarks-president-prepared-delivery-signing-stem-cell-executive-order-and-scientifi

Ombelet W, Van Robays J. Artificial insemination history: hurdles and milestones. Facts Views Vis Obgyn. 2015;7(2):137–43.

Oyez. Griswold v. Connecticut. Oyez Web site. https://www.oyez.org/cases/1964/496. Published 1965. Accessed 15 Feb 2021.

Oyez. Roe v. Wade. Oyez Web site. www.oyez.org/cases/1971/70-18. Published 1973. Accessed 15 Feb 2021.

Oyez. Planned Parenthood of Southeastern Pennsylvania v. Casey, 505 US 833. Oyez Web site. https://www.oyez.org/cases/1991/91-744. Published 1992. Accessed 15 Feb 2021.

Oyez. Gonzales v. Carhart, 550 US 124. Oyez Web site. www.oyez.org/cases/2006/05-380. Published 2007. Accessed 15 Feb 2021.

Pandia Health. Birth control and religion. Pandia Health Web site. www.pandiahealth.com/resources/birth-control-religion/. Published 2019. Accessed 15 Feb 2021.

Pernick MS. Taking better baby contests seriously. Am J Public Health. 2002;92(5):707–8.

Peter JC. Moral certitude in the use of levonorgestrel for the treatment of sexual assault survivors. In: Contemporary controversies in Catholic bioethics. Cham: Springer International Publishing; 2017. p. 197.

Petitti DB. Clinical practice: combination estrogen-progestin oral contraceptives. N Engl J Med. 2003;349(15):1443–50.

Planned Parenthood. How does the abortion pill work? Planned Parenthood Web site. www.plannedparenthood.org/learn/abortion/the-abortion-pill/how-does-the-abortion-pill-work. Published 2020. Accessed 15 Feb 2020.

Poston DL, Bouvier LF. Population and society: an introduction to demography. 2nd ed. New York: Cambridge University Press; 2017.

Rao M. Why all non-altruistic surrogacy should be banned. Econ Polit Wkly. 2012:15–7.

Rivera R, Yacobson I, Grimes D. The mechanism of action of hormonal contraceptives and intrauterine contraceptive devices. Am J Obstet Gynecol. 1999;181(5):1263–9.

RxList. Yasmin. 2020. www.rxlist.com/yasmin-drug.htm

Segall-Gutierrez P, Tilley I. Emergency contraception. Wiley-Blackwell; 2010.

Segers S, Pennings G, Dondorp W, de Wert G, Mertes H. In vitro gametogenesis and the creation of 'designer babies'. Camb Q Healthc Ethics. 2019;28(3):499–508.

Sermon K, Van Steirteghem A, Liebaers I. Preimplantation genetic diagnosis. Lancet. 2004;363(9421):1633–41.

Shimabukuro JO. Reviewing recently enacted state abortion laws and resulting litigation. 2019. https://pdfs.semanticscholar.org/d909/0e81bb237ebf7e5da8d5c70a9940494510c4.pdf

Sueoka K. Preimplantation genetic diagnosis: an update on current technologies and ethical considerations. Reprod Med Biol. 2016;15(2):69.

Sullivan D. The oral contraceptive as abortifacient: an analysis of the evidence. Perspect Sci Christ Faith. 2006;58(3)

Sullivan DM, Costerisan A. Complicity and stem cell research: countering the utilitarian argument. Ethics Med. 2008;24(3):151–8.

Taneja P. India's surrogate mother industry. BBC News. 2008. http://news.bbc.co.uk/2/hi/south_asia/7661127.stm. Published 11 Oct 2008.

The Emergency Contraception Web Site. Office of Population Research, Princeton University. https://ec.princeton.edu/questions/ecwork.html. Published 2020. Accessed 15 Feb 2021.

Thoma ME, McLain AC, Louis JF, et al. Prevalence of infertility in the United States as estimated by the current duration approach and a traditional constructed approach. Fertil Steril. 2013;99(5):1324–1331.e1321.

Thomson JJ. A defense of abortion. Philos Public Aff. 1971;1(1):47–66.

Tortora GJ, Derrickson B. Principles of anatomy & physiology. 15th ed. Hoboken: Wiley; 2018.

Treacy GC. Five great encyclicals. New York: The Paulist press; 1939.

Trussell J, Raymond E. Emergency contraception: a last chance to prevent unintended pregnancy. Princeton: Office of Population; 2016.

U.S. Supreme Court. Griswold v. Connecticut, 381 U.S. 479. U.S. Supreme Court. www.law.cornell.edu/supremecourt/text/381/479#ZO-381_US_479astref. Published 1965. Accessed 15 Feb 2021.

U.S. Supreme Court. Roe v. Wade, 410 U.S. 113. U.S. Supreme Court Web site. www.www.law.cornell.edu/supremecourt/text/410/113. Published 1973. Accessed 15 Feb 2021.

Verlinsky Y. Designing babies: what the future holds. Reprod BioMed Online (Reproductive Healthcare Limited). 2005;10:24–6.

von Hertzen H, Godfrey EM. Emergency contraception: the state of the art. Reprod Biomed Online. 2009;18(Suppl 1):28–31.

Wai Ngai S, Fan S, Li S, et al. A randomized trial to compare 24 h versus 12 h double dose regimen of levonorgestrel for emergency contraception. Hum Reprod. 2005;20(1):307–11.

Whalen K, Rose R. Ulipristal (Ella) for emergency contraception. Am Fam Physician. 2012;86(4):365–9.

Chapter 9
Ethics at the End of Life – Part I

Introduction

We turn now from the beginning of life and reproductive ethics to the other end of the spectrum, when physical life nears its conclusion. The care of terminal patients is often complicated and ethically challenging, as the focus of the healthcare interaction must necessarily change from cure to comfort. Yet even the provision of "comfort care" is a relatively new development in medicine, as the following vignette will reveal.

It was the summer of 1979. Writing now as one of the authors of this book (DMS), I was at the end of my first year of residency training in general surgery. At 6:30 in the evening, we were making post-op rounds at a university teaching hospital. The distinguished senior professor of surgery led the discussion at each bedside. In the room were the chief resident, two junior residents, two medical students, and myself, the first-year resident. Dr. Everett Grayson (not his real name) fired questions to his subordinates with rapid-fire efficiency:

"What's the story with Mrs. Johnson?"
"Uh, she's post-op day two after routine cholecystectomy for symptomatic gallstones."
"How's the wound? What's her white count? Has she been up out of bed?

Barely waiting for an answer to each query, the staff surgeon poked the poor woman's belly, said a few words to her, and moved on, with his entourage following after him. He led the group of white coats into the next room:

"What's up with Mr. Jasper?"
"Uh, sir, he was operated on yesterday for a left inguinal hernia."
"OK, let's look at the wound. Hmm, a little red – keep an eye on that, and let's get a white count. What's next?"

As we approached the next room, Dr. Grayson held up his hand, bringing the cluster of doctors to a halt:

> "I think I know this next case. Tell me about Mrs. Peterson."
> "Uh, Dr. Grayson, she's got pancreatic cancer, obvious stage four, open and shut abdominal exploration, no treatment."
> "Yeah, well, we don't need to see her."

And the medical team moved on to visit the next patient.

In our modern healthcare context, such insensitivity may seem like an anomaly, but at that time, it was all too common. Physicians focused on curing their patients and were uncomfortable with the dying, regarding them as failures. Symptomatic management of the dying was almost an afterthought, and reimbursement models focused primarily on curative or restorative treatments, discouraging clinicians from providing comfort care (Connor 2008).

Nonetheless, a new movement was taking place in the latter part of the twentieth century, led by two great pioneers. Psychiatrist Elisabeth Kűbler-Ross, in her 1969 seminal book *On Death and Dying*, described her interviews with dying patients and was the first to identify the five psychological stages that most of them go through: denial, anger, bargaining, depression, and acceptance (Kübler-Ross 1969). Her many lectures in academic and clinical institutions left a profound impression on students and practitioners and helped clinicians to become more comfortable talking with the terminally ill.

Even more influential was the great British nurse and physician Dame Cicely Saunders. One tribute described her legacy this way:

> Cicely Saunders founded the first modern hospice and, more than anybody else, was responsible for establishing the discipline and the culture of palliative care. She introduced effective pain management and insisted that dying people needed dignity, compassion, and respect, as well as rigorous scientific methodology in the testing of treatments. She abolished the prevailing ethic that patients should be cured, that those who could not be cured were a sign of failure, and that it was acceptable and even desirable to lie to them about their prognosis (Richmond 2005).

"Dame Cicely," as she was known, had deep compassion for those dying with cancer and felt that the healthcare system often ignored their needs. Her simple message was "constant pain needs constant control." She advocated for the provision of analgesics on a regular schedule, rather than only treating pain on-demand after it became severe. This plan not only improved mental health and comfort, but patients even lived longer as a result (Clark 2007). Today, palliative care is a rigorous medical specialty, and hospice care is readily available for patients where curative treatments are no longer possible.

Palliative care, simply defined, is treatment primarily aimed at symptoms, such as pain, breathlessness, or psychological distress related to a chronic illness. A multidisciplinary consulting team usually provides such treatment, with several specialty-trained palliative care physicians and nurses involved, along with social workers, dietary support, and chaplains. Contrary to widely-held misconceptions, palliative care is not just for the terminally ill, but for anyone with chronic debilitating symptoms. This type of management often provides curative treatments at the same time as palliative ones. By contrast, *hospice care* is aimed exclusively at symptom management, when curative treatments are no longer feasible or desired, and life expectancy is less than 6 months. Hospice can take place in many settings, including hospitals, long-term care facilities, and at home.[1] Pharmacists work closely with both types of specialties.

In discussing the ethics of the terminally ill, we turn now to the acute encounter between clinician and patient. Today's medical professional faces a unique dilemma: where is the balance between commitments to life and healing versus a willingness to "let go" when the time comes? We will explore this balance by using a case history as our point of departure. This approach will allow us to introduce and discuss some commonly misunderstood concepts related to end-of-life clinical management and to discuss the ethics involved.[2]

Case Study

Charles M., a 72-year-old retired accountant, presented to the emergency department of a community hospital in severe respiratory distress. He had a history of heavy tobacco use. Though he quit smoking altogether 2 years before this admission, he remained chronically short of breath and on supplemental home oxygen. Three days before admission, Charles began to notice an increase in his usual shortness of breath, a dry cough, and fever. On the day of admission, these symptoms grew worse, and he was brought to the emergency department by ambulance.

On physical exam, Charles was a thin, anxious, chronically ill appearing man in respiratory distress. He had a fever of 101.4 degrees F. Admission laboratory studies revealed normal serum electrolytes and blood hemoglobin, but a white blood count of 14,500 per cu. mm. His arterial blood gases were consistent with respiratory failure. A chest X-ray showed dense infiltrates in the right lower lobe.

In the emergency room, physicians orally intubated the patient and placed him on a mechanical ventilator. They admitted him to the medical intensive care unit with a diagnosis of chronic emphysema, with superimposed right lower lobe pneumonia and acute respiratory failure. Over the next several days, physicians treated Charles with antibiotics for his pneumonia. The lung infiltrates improved, and the patient's temperature and white blood cell count became normal. However, multiple attempts to wean him from the ventilator failed. Off the ventilator, he became restless and agitated, with severe shortness of breath.

---

[1] What Are Palliative Care and Hospice Care? National Institute on Aging Web site. www.nia.nih.gov/health/what-are-palliative-care-and-hospice-care#palliative. Published 2017. Accessed 9/17/2020

[2] Portions of this chapter first appeared in: Sullivan D. Euthanasia versus letting die: Christian decision-making in terminal patients. Ethics Med. 2005;21(2):109–18. Used with permission.

The patient's primary physician, a pulmonary specialist, discussed the various options with Charles and his family. All agreed that continued long-term reliance on the ventilator was burdensome and that his condition was terminal. The patient was fully alert and aware; he and his family understood the implications of his illness fully. A "do not resuscitate" (DNR) order was entered in the chart, with the full agreement of Charles and his wife. After a night of rest, the physician removed the endotracheal tube, and respiratory therapy took the ventilator from the room. The patient remained on supplemental oxygen. Ten hours after discontinuing ventilator support, and with his family present, the patient died.

This example comes from an actual case witnessed by one of the authors (DMS). We do not intend for the case to be controversial, but it may raise several questions. For example, was the cessation of therapy for Charles justified? What ethical principles guided the clinicians?

Our discussion will benefit from a careful definition of terms. A *terminal condition* is a disease or process that will eventually result in a patient's death, no matter the treatment given. Of course, this may include cases where death is inevitable but far off, as in patients with cancer who live for a few years with their disease. On the other hand, the expression *imminent death* means that death will occur within a short time, usually days or weeks (Kilner 1992; Goldsteen et al. 2006).

With these definitions in mind, was the cessation of therapy for Charles justified? The patient's condition was terminal, and his death was imminent. There should be no reason to second-guess the physician's judgment here. The patient was in respiratory failure and dependent on the ventilator because of end-stage lung disease from chronic emphysema. He had received the best of aggressive medical therapy, and further such treatments might be considered medically futile.

It is worth noting that medical personnel may abuse the concept of *futility*, often on arbitrary or utilitarian grounds. For example, treatment may be withdrawn because of a vague perception that the patient lacks quality of life. "Quality of life" is a subjective measure of personal satisfaction expressed by individuals about their physical, mental, and social situation, and is based on autonomy. However, the phrase is fraught with possible misunderstanding. The judgment of a poor quality of life should best be made by the one living that life or the patient's family. Human beings are amazingly resilient and may be quite content with an experience that outside observers may deem intolerable. Therefore, clinicians should be extremely cautious about making quality of life determinations for their patients (Jonsen et al. 2015).

The case of Charles M., however, is an example of the best kind of clinical relationship. Out of respect for the patient's dignity and aware of his dire medical condition, the healthcare team communicated openly with Charles and his family. Full and informed consent was sought and given by all parties. It is important to note that the patient could have chosen to remain ventilator-dependent for an extended time. Such a plan would have required a tracheostomy and transfer to a long-term care facility. But he had found the machine intrusive and burdensome; he simply did not wish to live that way.

One reason that this case may seem challenging is that the doctors withdrew an already utilized treatment (the ventilator) as opposed to withholding it. Some might argue that the doctors in the emergency room should never have intubated Charles and placed him on a ventilator in the first place, yet this would have been a denial of any attempt to treat him, and inappropriate. Having established that further ventilator support was futile, the decision to withdraw it seems justified. In the past, many have made a distinction between *withholding* treatment, i.e., not starting it, versus *withdrawing* treatment, i.e., stopping an intervention already begun. Historically, the latter has always been more difficult in clinical practice than the former, though this is probably more psychological than real. Beauchamp and Childress have said that "the distinction between withholding and withdrawing is morally untenable and can be morally dangerous." (Beauchamp and Childress 2019).

A helpful guide, in this instance, is the *principle of double effect*. This concept gives significant weight to intentions in moral decision-making. For example, caregivers are obligated to preserve life and, at the same time, to relieve pain. If a physician were to inject a massive overdose of morphine into a terminally ill cancer patient, with the intent of active euthanasia, this would be morally wrong, according to most clinicians. Most would also condemn a pharmacist who compounded this or a similar lethal injection on a physician's orders. However, the healthcare team should endeavor to treat the pain of a suffering patient with adequate doses of analgesics, including narcotics. This principle assumes that other medications have failed, and that imminent death makes drug dependency irrelevant. If such treatment somewhat hastens the death of the patient, but this was an *unintended* consequence of the intent to relieve suffering, then the act may be morally permissible (Noia 2017). Applying this idea to the case of Charles M., neither the patient nor the clinicians intended his death. They did, however, intend to relieve him from a burdensome and futile treatment (the ventilator); his death, therefore, was the unintended consequence of a beneficent intent. The principle of double effect gives justification for this action.

As we conclude our discussion of Charles M., it should be clear that all of the concepts of medical principlism have been operative: autonomy, beneficence, nonmaleficence, and distributive justice. While challenging, the management of this case was not ethically very controversial. However, a great deal of ethical debate arises with our next major topic, the contentious yet fascinating arena of defining death and near-death.

Defining Death

It was 3:05 a.m., and family medicine resident Dr. Helen Jamieson had just completed admitting a patient from the emergency department. He was an older man with a relatively routine case of pneumonia. After admitting the patient and starting him on antibiotics, Helen went up to the on-call room, hoping to get a few precious hours of sleep. Her head had just hit the pillow when her cell phone went off. The apologetic nurse on the line told her that the patient in 521 had stopped breathing; could Dr. Jamieson come up and "pronounce" her? Helen groaned as she grabbed her stethoscope and penlight and headed up to the fifth floor.

The yawning resident examined the still form in room 521. The elderly patient had been in a comatose state for several days after a long battle with metastatic breast cancer. She was now motionless on the bed, with cold extremities and no visible breathing. Helen listened to her lungs with her stethoscope; she could hear no breath sounds and could detect no attempts at respiration. Shifting her attention to the heart region, she listened carefully for heart sounds but could detect none, and there was no detectable pulse. Helen then took her penlight and checked for pupillary responses, but the pupils were fixed and dilated, with no reaction to light. Finally, she pushed hard on the patient's right supraorbital ridge but could elicit no response to pain.[3] Helen left the room and went over to the nursing station, where the paperwork was waiting for her. The doctor signed the death certificate, "pronouncing" her dead at 3:52 a.m. She sighed as she went back to the on-call room, hoping to get a bit of rest at last.

Almost every physician in training has had to perform some variant of the duties described in this case, and pharmacists should be conversant with these details as well. Indeed, the diagnosis of cardiopulmonary death is not difficult, for the signs are readily apparent. This approach is the standard of death applied to the overwhelming majority of patients since the advent of medicine.

However, by the middle of the twentieth century, advances in positive pressure ventilation and the resuscitation of critically ill patients led to the perceived need for a more flexible definition of death. In 1968, anesthesiologist Henry Beecher chaired a Harvard committee designed to examine the problematic and vexing problem of patients in an irreversible coma. At the time, it was considered ethically mandatory to continue to hydrate and ventilate such patients, even though they had no hope of survival. Their proposed definition of "brain death," as a new definition of death, appeared in the *Journal of the American Medical Association* in August of 1968 (A Definition of Irreversible Coma 1968). The new formulation was widely criticized as an invention, meant solely to facilitate organ transplantation, even though that was never the intent of the original report (DeGeorgia 2014).

During the 1960s, organ transplantation was becoming more feasible but always guided by the *Dead Donor Rule*, which states that the retrieval of organs for transplantation should not cause the death of a donor. This concept was the ethical foundation for the use of cadaver organs as transplants (Truog and Miller 2008). The new approach of brain death conveniently allowed patients to be declared dead while still maintaining perfusion of the kidneys, the liver, and even the heart and lungs, and a new era of transplantation began. In 1981, the President's Medical

---

[3] Jones O. How to Certify Death. *Teach Me Surgery.* 2017. https://teachmesurgery.com/examinations/misc/confirmation-of-death/

Ethics Commission drafted a model for legislation with the following criteria: An individual with (1) irreversible cessation of circulatory and respiratory functions, or (2) irreversible cessation of all functions of the entire brain, including the brain stem, is dead. Over the next several years, all fifty U.S. states and the District of Columbia enacted this "whole brain" definition of death by statute or judicial review (Jonsen et al. 2015; Capron 1988).

In practice, the following steps are required to establish a diagnosis of brain death, better referred to as death by neurological criteria:

- No movement and no reflexes, e.g., no reaction to painful stimuli, no pupillary constriction to light, and no deep tendon reflexes.
- No spontaneous breathing, as determined by a positive *apnea test*, where the patient's ventilator is temporarily disconnected with no evidence of breathing effort.
- Additional diagnostic tests may be confirmatory, such as demonstrating an isoelectric EEG or the absence of cerebral blood flow by radionuclide scan.
- All of this presupposes no narcotic drugs in the bloodstream and a normal body temperature (Jonsen et al. 2015; Scott et al. 2013).

When these criteria have been met, it is both legal and ethical to remove transplantable organs (assuming appropriate permissions have been obtained) before discontinuing the ventilator and intravenous fluids, and this respects the Dead Donor Rule.

Dead Donor Rule

The retrieval of organs for transplantation should not cause the death of a donor.

Though most religious authorities have accepted brain death, there remain some nagging philosophical and theological concerns that we cannot address in this brief review. Some conservative Roman Catholic voices, for example, have questioned the concept, even though papal authorities have endorsed the idea, along with organ transplants made possible by it.[4]

Nonetheless, these standards have been almost universally implemented in ethics and law. In clinical practice, where organ donation is not feasible or consented to, further medical interventions are unnecessary in brain dead patients. Physicians are legally permitted to declare death, and there is no requirement to seek permission from the family before discontinuing the ventilator and other treatments (Jonsen et al. 2015) (Fig. 9.1).

---

[4] Allen JL. Vatican newspaper article challenges 'brain death' notion. *National Catholic Reporter.* 2008. www.ncronline.org/news/vatican-newspaper-article-challenges-brain-death-notion-0. Published 9/3/2008

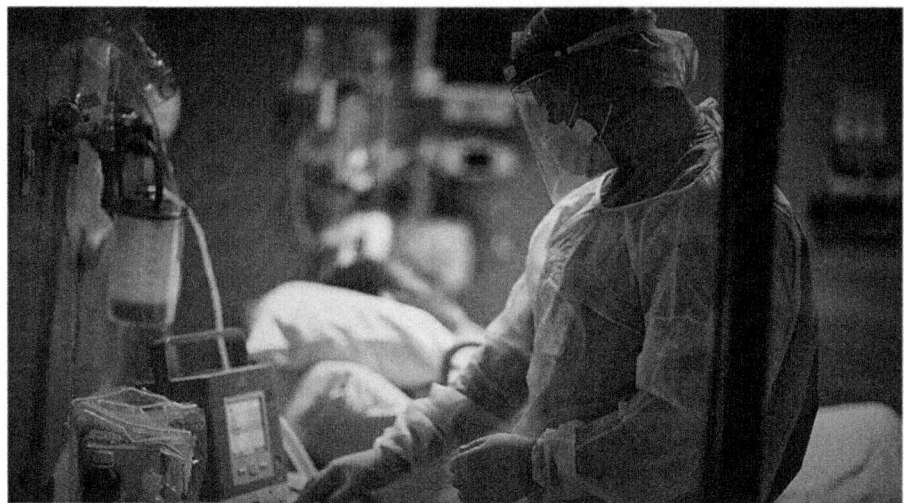

Fig. 9.1 Patient connected to a ventilator during an ICU night shift at Baton Rouge General Hospital (Lt. Cmdr. Michael Heimes, assigned to Expeditionary Medical Facility-M, checks on a patient connected to a ventilator during an ICU night shift at Baton Rouge General Mid City campus, April 28, 2020. (U.S. Marine Corps photo by Cpl. Daniel R. Betancourt Jr./Released) 200428-M-WU117-1661)

The Persistent Vegetative State

Despite the near-universal acceptance of brain death criteria, there is much more ethical uncertainty surrounding a diagnosis of a *persistent vegetative state* (PVS), a clinical scenario where the patient has suffered severe cerebral damage, but the brainstem is intact. Such cases feature a state of deep unresponsiveness with preserved sleep/wake cycles. Reflexes and spontaneous breathing are preserved. Such patients do not meet the criteria for brain death and are not dependent on a ventilator. They need only tube feedings and hydration, and may thereby be kept alive indefinitely (Jonsen et al. 2015). Is it ethical to discontinue artificial nutrition and hydration and allow such severely handicapped patients to die?
PVS cases have been the occasion for a great deal of legal scrutiny. We will briefly mention three of the most significant "right to die" precedents in law. 21-year-old Karen Ann Quinlan suffered irreversible brain damage in April of 1975, after complications from a crash diet coupled with valium and alcohol use. Her parents wanted to have her ventilator removed, which the hospital resisted. Multiple court battles eventually ended up in the New Jersey Supreme Court, which ruled in March of 1976 that the ventilator could be discontinued. Surprisingly, when the ventilator was removed in May of that year, she survived. Afterward, she was kept alive with tube feedings until her death from pneumonia in 1985 (McFadden 1985).

In 1983, 25-year-old Nancy Cruzan suffered a nighttime car accident in which she was thrown from her car. Paramedics resuscitated her, but she was eventually

> **Brain Death versus PVS**
>
> **Brain Death:**
>
> - No movement or reflexes
> - No spontaneous breathing (failed apnea test)
> - Isoelectric EEG or no cerebral blood flow by scan (confirmatory).
> - Negative narcotic screen, normal body temperature.
>
> **Persistent Vegetative State:**
>
> - Deep unresponsiveness
> - Preserved sleep/wake cycles
> - Not ventilator dependent
> - Tube feedings and I.V. hydration required

diagnosed with a PVS and had a feeding tube placed for long-term nutrition and hydration. Nancy's parents requested that the hospital remove artificial nutrition and hydration, which the hospital was unwilling to do without court authorization. Despite approval by a lower court, both the state Supreme Court and the U.S. Supreme Court eventually ruled against the parent's wishes, stating that the interest of the state to preserve life outweighed the privacy interests of Nancy and her family. The high court's 5–4 ruling was not the end of the story, however. In December of 1990, several of Nancy's friends testified before a Missouri probate judge, who ruled that there was "clear and convincing" evidence that she would have refused further life-sustaining treatments. The hospital removed her feeding tube, and she died 12 days later (Stonecipher 2006).

The most recent and undoubtedly best-known "right to die" case is that of Terri Schiavo. In February of 1990, 26-year-old Terri suffered significant brain damage after cardiac arrest, probably caused by a potassium imbalance related to an eating disorder. She was diagnosed with a PVS and was placed in a skilled nursing home for ongoing tube feedings. After 8 years of caring for her, her husband asked a Florida court to authorize the removal of her feeding tube. Terri's parents resisted this, refusing to accept the PVS diagnosis. There ensued a prolonged and very public legal and political battle, eventuating in the Florida legislature passing "Terri's Law," designed to force the hospital to keep Terri's feeding tube in place, but declared unconstitutional by the Florida Supreme Court. Attempts to counter this decision before the 11th Circuit Court of Appeals were unsuccessful. On two occasions, the U.S. Supreme Court declined to hear aspects of the case. On March 31, 2005, Terri Schiavo died at age 41, 13 days after her feeding tube was removed for the last time (Stonecipher 2006; Hook and Mueller 2005).

What ethical insights can we gain by studying these contentious cases? One lesson is the distinction between so-called "ordinary" and "extraordinary" treatments. Traditional morality, primarily that endorsed by the Roman Catholic Church, has taught that some treatments are "ordinary," e.g., food, water, and (perhaps) antibiotics. On this view, such treatments are so essential and minimally invasive as to be mandatory in all cases. Nutrition is nurture, and therefore never optional. For example, in 2004, Pope John Paul II made a strong statement on the care of patients in a PVS, declaring that the provision of tube feedings was always obligatory (Barry 2004). It is worth noting that this understanding is closely related to Catholic views on natural law, as we discussed back in Chap. 2.

Contrast this with "extraordinary" or "heroic" treatments, such as a ventilator, renal dialysis, or a heart transplant. These highly invasive and technological medical interventions should only be used when the patient's condition warrants it and may be declined when death is near (Sullivan 2007). However, some reflection may reveal a problem with this approach: A ventilator may seem "extraordinary" in a patient with end-stage lung failure (such as Charles in our earlier case study). However, it is "ordinary," indeed essential, in a young, healthy patient with a crush injury to the chest, who depends on the ventilator to stay alive until the chest wall heals and she can breathe on her own.

In recent years, as ethics language has shifted and medical care has gotten more complex, the ordinary/extraordinary distinction may not be as helpful. Instead, the language of proportionality is employed, with the emphasis on the patient rather than the treatment: some treatments are proportionate to reasonable treatment goals, and some are not. With this in mind, tube feedings, though a "minimum" intervention technologically, may still be disproportionate if the prognosis is poor (Orr 2004). Despite many years of lively debate, controversies over artificial nutrition and hydration in the desperately ill have continued until the present day. Unlike cases involving brain death, patients with a diagnosis of PVS still present a vexing challenge to clinical ethicists.

Advance Directives

Despite a divisive and often bitter debate over recent "right to die" cases, one outcome has been very positive. These controversies have spurred greater public interest in *advance directives*. These legal documents allow patients and their families to make their end-of-life wishes known long before severe illness strikes. The two most common advance directives are a living will and a durable power of attorney for health care (Fig. 9.2).

A *living will* is a written statement of a patient's treatment preferences should he or she become incapacitated with a life-threatening illness. Living wills are recognized as a legal expression of a patient's autonomous wishes in most states, usually only requiring the signature of two witnesses and sometimes that of a notary. In some states, a living will is invalid if the patient is pregnant.[5] One example of typical language in a living will is shown here:

---

[5] Living Wills: State Laws. FindLaw Web site. https://estate.findlaw.com/living-will/living-wills-state-laws.html. Published 2020. Accessed 12/18/2020.

Advance Directives

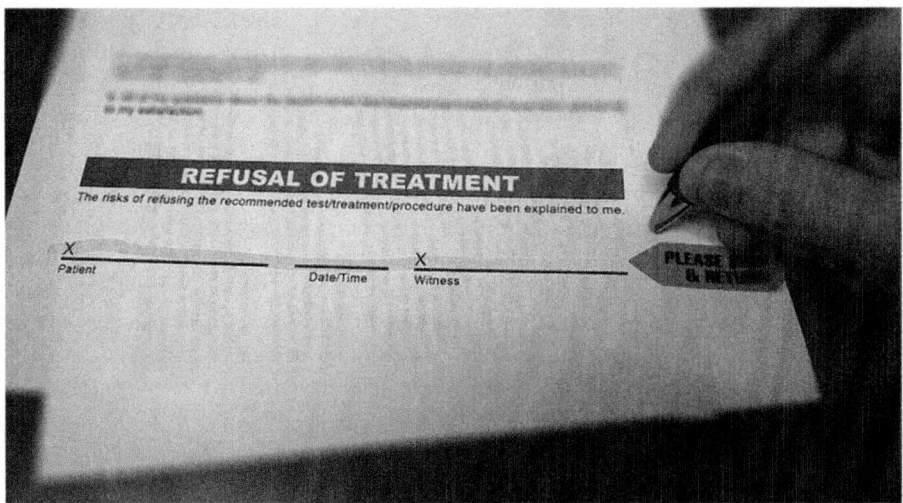

Fig. 9.2 Generic refusal of treatment form awaiting signature. (Illustration in Public Domain: Generic refusal of treatment form awaiting signature)

> I direct my attending physician or primary care physician to withhold or withdraw life-sustaining medical care and treatment that is serving only to prolong the process of my dying if I should be in an incurable or irreversible mental or physical condition with no reasonable medical expectation of recovery.[6]

The document may then further specify specific treatments that the patient would like to refuse if incapacitated and terminally ill, such as a ventilator, dialysis, or CPR. A patient must have *decision-making capacity* (discussed in Chap. 4) to create a living will. There are several potential problems with the implementation of living wills:

- Patients and their families must make the signed copy of the document available to the healthcare provider at the time of treatment. In a rush to get a patient to the hospital in a medical emergency, family members may not always have this with them.
- The provisions of a living will can only be carried out if the patient currently lacks decisional capacity and has a life-threatening incurable illness.
- Living wills often use vague, imprecise terms, such as "heroic measures," leaving providers in doubt as to the patient's intent.
- Patient preferences are not always followed, even when a living will is in place.
- Living wills are inflexible. Often written years before they are invoked, they may not anticipate newer treatments that may be in the patient's actual best interest. There is no provision for providers to override the original language of the document.

---

[6] Sample Living Will Form. FindLaw Web site. https://estate.findlaw.com/living-will/sample-living-will-form.html. Published 2020. Accessed 12/18/2020.

- A living will is not itself a medical order. Certain providers, such as paramedics, are in a bind when it comes to implementing them, and may legally be forced to provide treatments that the patient would not want (Beauchamp and Childress 2019; Higel et al. 2019; Devettere 2016; Levenson and Zucker 2017).

To mitigate some of these concerns, many ethicists recommend the use of a *durable power of attorney for healthcare* (DPAHC). Also known as a *healthcare proxy*, the patient designates a family member or friend as their representative for healthcare decisions. Note that this document is not the same as a legal power of attorney for financial and other life decisions; it only applies to the provision of clinical treatments. With a DPAHC, the patient's representative can make decisions on behalf of their loved one, and can take into account current state-of-the-art remedies that may be helpful or at least worth implementing on a trial basis. Possible limitations of a DPAHC include:

- Once again, patients and their families must make the signed copy of the document to the healthcare provider at the time of treatment. If the patient lacks decisional capacity, and the document is not available, the "wrong" proxy might be forced to make treatment decisions.
- Some designated proxies may be uncomfortable with the weight of treatment decisions. They may not wish to experience the guilt of "killing" their loved one by foregoing an ineffective treatment.
- The designated proxy may not be available.
- As with living wills, a DPAHC is not itself a medical order, leaving some types of clinicians without the authority to follow it (Jonsen et al. 2015; Devettere 2016; Levenson and Zucker 2017).

Some excellent advance directives provide provisional language for the designated proxy to guide them as to their desires to forgo CPR or a ventilator, for example. However, if both a living will and a DPAHC are in place, the provisions of the living will *supersede* the DPAHC. A healthcare proxy cannot overrule a living will.[7] By way of context, if a living will had been in place in the case of Terri Schiavo, her case would never have gone to court, and the feeding tube would have been removed when this was first proposed.

Finally, we will make a brief mention of *Physician Orders for Life-Sustaining Treatments*, or POLST. Sometimes referred to as MOLST (Medical Orders for Life-Sustaining Treatments), most states have made provision for this type of directive, which offers the following features:

- Patients can fill out a POLST form when they have decisional capacity and a life expectancy of less than a year.
- POLST forms reflect a patient's preferences, translating them directly into medical orders.

---

[7] The Power of Attorney, Living Will, and Your Healthcare. FindLaw Web site. https://estate.findlaw.com/living-will/the-power-of-attorney-living-will-and-your-healthcare.html. Published 2020. Accessed 12/18/2020

- Unlike other advance directives, POLST forms become an actionable set of medical orders that transfer across institutions (e.g., from nursing homes to hospitals). This includes direct orders for paramedics and nurses (Devettere 2016; Levenson and Zucker 2017)[8].

Despite some limitations, most clinical authorities feel that advance directives are a good idea. One recent analysis showed that about 37% of Americans had such a document, with only 50% of those surveyed currently suffering from a chronic illness.[9]

Conclusion

This chapter has been a deep dive into the ethics surrounding end-of-life clinical decisions. We looked briefly at the history of clinical care for the dying, which only recently has matured to the point of allowing for palliative care and hospice. We then used a case study to examine the many aspects of caring for a terminally-ill patient, revealing the competing duties informing the decision to withdraw a futile treatment. Our analysis then moved to an examination of two definitions of death, one cardiorespiratory and the other neurological. We then considered the vexing clinical problem of the persistent vegetative state, at the heart of three controversial "right-to-die" legal cases. We then examined the matter of advance directives. The following chapter will continue our discussion of ethics at the end of life, looking into the highly controversial areas of assisted suicide and pharmacist participation in lethal injections for capital punishment.

Key Terms

- palliative care v. hospice
- terminal condition v. imminent death
- medical futility
- quality of life
- withholding v. withdrawing treatments
- principle of double effect
- cardiopulmonary death v. brain death
- dead donor rule
- persistent vegetative state
- ordinary v. extraordinary treatments
- proportionate v. disproportionate treatments

---

[8] Carley M, Maag M, Pope TM, Sabatino CP, Vandenbroucke A, Wolf R. POLST legislative guide. *National POLST Paradigm Task Force.* 2014

[9] Crist C. Over one third of U.S. adults have advanced medical directives. *Reuters News Service.* July 11, 2017, 2017. www.reuters.com/article/us-health-usa-advance-directives/over-one-third-of-u-s-adults-have-advanced-medical-directives-idUSKBN19W2NO

- living will
- durable power of attorney for health care (or healthcare proxy)
- POLST

Questions for Review

1. Discuss the ethical justification for removing the ventilator in the case of Charles M.
2. Is palliative care only for dying patients? In what way does this differ from hospice?
3. How does *intent* help determine the ethics of a treatment with possible adverse effects?
4. What are the clinical criteria used to justify a diagnosis of brain death? When brain death is established, can family members oppose the cessation of treatments?
5. What unique features constitute the persistent vegetative state? Is it ethical to withdraw treatment from such patients?
6. Discuss the relative strengths and weaknesses of various advance directives, including a living will, a durable power of attorney for health care, and POLST.

For Further Reading

Atul Gawande. *Being Mortal: Medicine and What Matters in the End*. Picador, 2017. A surprisingly readable and deeply moving account, written by a surgeon. Gawande reveals the often disturbing facts about how Americans die, and suggests how we might think about our inevitable mortality.

References

A Definition of Irreversible Coma: Report of the Ad Hoc Committee of the Harvard Medical School to Examine the Definition of Brain Death. J. Am. Med. Assoc. 1968;205(6):337–40.

Barry R. The papal allocution on caring for persons in a vegetative state. Issues Law Med. 2004;20:155.

Beauchamp TL, Childress JF. Principles of biomedical ethics. 8th ed. New York: Oxford University Press; 2019.

Capron AM. The report of the President's commission on the uniform determination of death ACT. In: Zaner RM, editor. Death: beyond whole-brain criteria. Dordrecht: Springer Netherlands; 1988. p. 147–70.

Clark D. From margins to centre: A review of the history of palliative care in cancer. Lancet Oncol. 2007;8(5):430–8.

Connor SR. Development of hospice and palliative care in the United States. OMEGA J. Death Dying. 2008;56(1):89–99.

DeGeorgia MA. History of brain death as death: 1968 to the present. J. Crit. Care. 2014;29(4):673–8.

Devettere RJ. Practical decision making in health care ethics: cases, concepts, and the virtue of prudence. 4th ed. Washington, DC: Georgetown University Press; 2016.

Goldsteen M, Houtepen R, Proot IM, Abu-Saad HH, Spreeuwenberg C, Widdershoven G. What is a good death? Terminally ill patients dealing with normative expectations around death and dying. Patient Educ. Couns. 2006;64(1–3):378–86.

References

Higel T, Alaoui A, Bouton C, Fournier JP. Effect of living wills on end-of-life care: a systematic review. J. Am. Geriatr. Soc. 2019;67(1):164–71.

Hook CC, Mueller PS. The Terri Schiavo Saga: the making of a tragedy and lessons learned. Mayo Clin. Proc. 2005;80(11):1449–60.

Jonsen AR, Siegler M, Winslade WJ. Clinical ethics: a practical approach to ethical decisions in clinical medicine. 8th ed. New York: McGraw-Hill Education; 2015.

Kilner JF. Life on the line. Bannockburn: Center For Bioethics and Human Dignity; 1992.

Kübler-Ross E. On death and dying. New York: Macmillan; 1969.

Levenson JD, Zucker K. The practical distinction between living wills and physician (Pennsylvania) orders for life-sustaining treatment (POLST). Proceedings of the annual meetings of the ACLM. J. Leg. Med. 2017;37:1–3.

McFadden RD. Karen Ann Quinlan, 31, dies; focus of right to die case. N. Y. Times. 1985;27(12):A1.

Noia GM. The principle of double effect within Catholic moral theology: a response to two criticisms of the principle in relation to palliative sedation. J. Moral Theol. 2017;6(2):130–48.

Orr RD. Ethics & life's ending: an exchange. First Things Mon. J. Relig. Public Life. 2004;145:31–6.

Richmond C. Dame Cicely Saunders, founder of the modern hospice movement. Br. Med. J. 2005. www.bmj.com/content/suppl/2005/07/18/331.7509.DC1

Scott JB, Gentile MA, Bennett SN, Couture M, MacIntyre NR. Apnea testing during brain death assessment: a review of clinical practice and published literature. Respir. Care. 2013;58(3):532–8.

Stonecipher M. The evolution of surrogates' right to terminate life-sustaining treatment. AMA J. Ethics. 2006;8(9):593–8.

Sullivan D. Euthanasia versus letting die: Christian decision-making in terminal patients. Ethics Med. 2005;21(2):109–18.

Sullivan SM. The development and nature of the ordinary/extraordinary means distinction in the Roman Catholic tradition. Bioethics. 2007;21(7):386–97.

Truog RD, Miller FG. The dead donor rule and organ transplantation. N. Engl. J. Med. 2008;359(7):674–5.

Chapter 10
Ethics at the End of Life – Part II

Our last chapter focused on time-honored ethical standards for ethics at the end of life. We now turn to some much more controversial concerns, and these are closely related to pharmacy practice.

The Assisted Suicide Debate

Among the traditional rules for healthcare ethics, personal autonomy has emerged as the dominant factor in recent discussions.[1] While patients have long had the right to decline life-prolonging treatments, there is a growing trend towards permitting suffering patients to choose the time and manner of their death and to have medical assistance in doing so. As a result, our society is having serious conversations about making euthanasia and assisted suicide acceptable as medical care for the terminally ill.

The word *euthanasia* comes from two Greek roots: *eu* for "good" and *thanatos* for "death." The term, therefore, simply means a "good" or "gentle" death.[2] In practice, euthanasia is the active intent to cause death as a form of medical treatment, sometimes referred to as "mercy killing." Though now legal in Canada and several other countries, euthanasia remains illegal throughout the U.S. *Physician-assisted suicide* is a variation of the intent for the patient to die, where the agent that causes the death is the patient herself, with means provided by clinicians (Devettere 2016). By contrast with euthanasia, assisted suicide has now become legal in many U.S. states. Therefore, our analysis will primarily focus on assisted suicide, though euthanasia will sometimes come up during the discussion.

---

[1] Portions of this chapter first appeared in: Sullivan DM, Taylor RM. The ethical landscape of assisted suicide: a balanced analysis. Ethics Med. 2018;34(1):49–57. Used with permission.

[2] Euthanasia. Vocabularycom. 2020. www.vocabulary.com/dictionary/euthanasia

Until recently, the American Medical Association has opposed assisted suicide in its Code of Ethics, claiming it is "fundamentally incompatible with the physician's role as healer."[3] In 2016, a motion was made to amend this statement to "neutral," but this was defeated, and the AMA remains in opposition (Van Way 2019). Many other healthcare societies still oppose the practice, including the National Hospice and Palliative Care Organization, the American College of Physicians, the American Academy of Pediatrics, the American Nurses Association, and the World Medical Association (Sulmasy et al. 2018). The American Pharmacists Association (APhA) is neutral about assisted suicide, stating that it is a matter of individual discretion (Hughes 2017; Fass and Fass 2011). Similarly, the American Society of Health-System Pharmacists (ASHP) has taken a position of neutrality, endorsing "the right of a pharmacist to participate or not in morally, religiously, or ethically troubling therapies." (Hughes 2017)

We will now examine the legal and ethical terrain of medically-assisted death, with an evaluation of both sides of this contentious issue. The subject of assisted suicide is controversial and contested, but a robust discussion requires a dispassionate examination of all the ethical arguments. This section will, therefore, present the case for and against this practice, with our analysis excluding strictly religious arguments. One important caveat is in order: For clarity, we will use the older, more descriptive term, "physician-assisted suicide" (PAS), instead of the newer terms "medical aid in dying" or "physician aid in dying." However, we do not intend for this choice of terminology to have any moral implication.

Current Legal Context in the United States

As of early 2021, eight states and the District of Columbia have legalized PAS. In chronological order, they are Oregon (1994), Washington (2008), Vermont (2013), California (2015), and Colorado (2016). The District of Columbia passed its law in 2016. Hawaii was added in 2018, followed by New Jersey and Maine in 2019.

States with Legal Assisted Suicide

Oregon
Washington
Vermont
California
Colorado
District of Columbia
Hawaii
New Jersey
Maine
Montana (by court ruling)

---

[3] AMA Code of Medical Ethics. American Medical Association Web site. www.ama-assn.org/delivering-care/ama-code-medical-ethics. Published 2017. Accessed 12/19/2020.

Current Legal Context in the United States

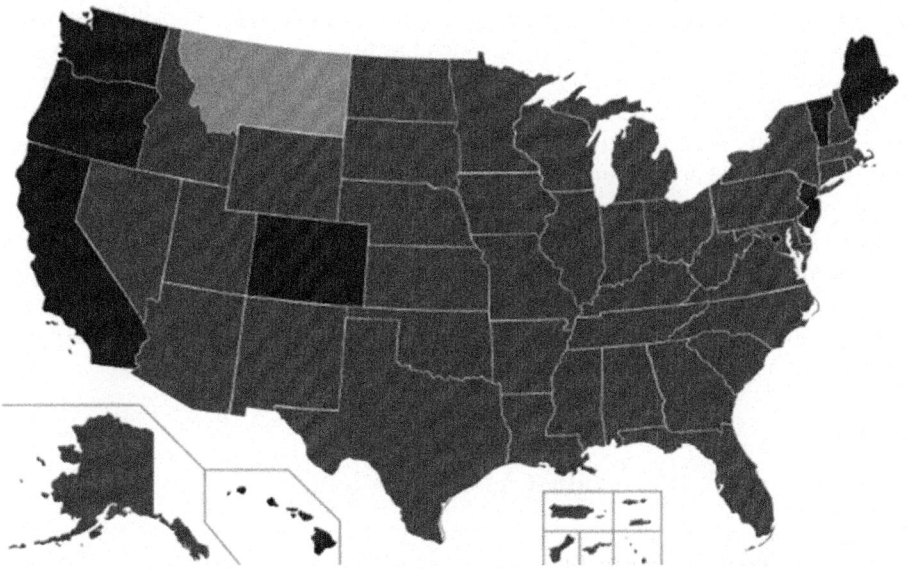

Fig. 10.1 As of 2021, states with PAS laws (dark grey). One state with PAS approval by court ruling (light grey). (File:Legality of Assisted Suicide in the US.svg. CC BY-SA 4.0)

Montana is unusual in having a State Supreme Court ruling (2009) that shields physicians from liability for assisting with suicide, but that state has no governing statute. At present, several additional states are actively considering the passage of "death with dignity" laws (Fig. 10.1)[4].

Typical requirements for PAS laws include the following:

- Eligible Patients: At least 18 years old, with established state residency and decisional capacity. Must have a terminal illness, with death anticipated within 6 months.
- Participating Doctors: Must be licensed to practice in the state. A mental health assessment is not mandatory. The physician must inform the patient of other possible options and must notify the patient's next-of-kin.
- Timing: Initial verbal request, followed by a 15-day waiting period. There must then be a second verbal request to the physician, followed by a written request. Finally, there must be a 48-hour waiting period before the patient can pick up the medications from a participating pharmacy.[5]

There are many background reasons for the ascendancy of this debate in our modern discourse. Several writers have commented on the rise of medical

---

[4] State-by-State Guide to Physician-Assisted Suicide. ProCon Web site. http://euthanasia.procon.org/. Published 2020. Accessed 2/15/2020.

[5] How Death with Dignity Laws Work. www.deathwithdignity.org/learn/access/. Published 2020. Accessed 12/19/2020.

technologies in the past half-century with its associated technological imperative. Consider hemodialysis, for example, which has been shown to increase life expectancy while not necessarily improving quality of life. Clinicians have traditionally been reluctant to limit such treatments, even in cases where the patient has dementia or is terminally ill (Swetz et al. 2012). At the same time, a rise in personal autonomy has led many to feel uncomfortable with the prospect of "prolonging the dying process" and a desire to exercise more control over personal decisions at the end of life. The older era of paternalism in healthcare has been replaced by an emphasis on respecting patient autonomy and personal values. PAS proponent Derek Humphry has summarized this succinctly, stating, "… every competent adult has the incontestable right to humankind's ultimate civil and personal liberty – the right to die in a manner and at a time of their own choosing."[6]

Ethical Arguments Against Assisted Suicide

We begin with arguments opposing PAS since this is the traditional view endorsed for over two millennia. Prohibitions against the direct taking of life are deeply embedded in our religious and legal codes. The Hippocratic tradition of non-maleficence is at the heart of this, as we have discussed in Chap. 4 (Edelstein 1943; Bulger and Barbato 2000). The idea of non-maleficence is also apparent in the Oath and Prayer of Maimonides, dating back to the twelfth century.[7] Despite, or perhaps because of its ancient pedigree, non-maleficence remains relevant to contemporary patients and clinicians. We have emphasized its application throughout this text (Fig. 10.2).

As we have also discussed previously, German philosopher Immanuel Kant taught an ethical theory based on two categorical imperatives (see Chap. 2). These can be summarized as follows:

1. Act according to a maxim that could become a universal law.
2. Act so that you always treat others as an end, never as a means to an end.[8]

This deontological (principles-based) approach militates against suicide in any context, which violates both categorical imperatives. It is not morally generalizable, and Kant held that suicide violates the end-versus-means distinction. However, Kant's approach is so deontological that it admits of no exceptions. Therefore, it cannot handle conflicting moral duties, so we will need a more robust approach to deal with patients suffering at the end of life.

---

[6] Humphry D. Liberty and death: a manifesto concerning an individual's right to choose to die. Assisted Suicide Web site. www.assistedsuicide.org/liberty_and_death_manifesto_right_to_die.html. Published 2009. Accessed 12/19/2020.

[7] Oath and Prayer of Maimonides. John Hopkins Sheridan Libraries Web site. https://guides.library.jhu.edu/bioethics/codes. Published 2017. Accessed 12/19/2020.

[8] Johnson R, Cureton A. Kant's moral philosophy. Stanford Encyclopedia of Philosophy. 2016. https://plato.stanford.edu/entries/kant-moral/

Fig. 10.2 Hippocrates. Line engraving by Desrochers after P. P. Rubens. (Credit: Hippocrates. Line engraving by Desrochers after P. P. Rubens. Credit: Wellcome Collection. Attribution 4.0 International (CC BY 4.0).)

In fact, at the heart of the moral case against PAS is the conflict of moral duties generated by medical principlism, which emphasizes non-maleficence, but also beneficence, distributive justice, and autonomy (recall that Kant was also the father of the concept of autonomy). As we discussed in Chap. 7, the duty of beneficence is a covenantal relationship, a more profound commitment than merely a promise or a contract. On this understanding, PAS erodes the trust implied by the healthcare relationship. Specifically, vulnerable patients place themselves in the care of clinicians with the expectation that they will act in their best interests.

We must also acknowledge that suffering is an unavoidable part of serious illness, and addressing it is a fundamental responsibility of healthcare (Cassell 2004). Most suffering related to illness can be attenuated and managed by appropriate interventions, whether medical, psychosocial, existential, or spiritual (Cherny et al. 2015). Indeed, unmanageable suffering is rare at the end of life when proper resources and interventions are provided. This is not to say that patients never complain of unbearable suffering, only to argue that, when appropriate services are provided, such patients are uncommon, as demonstrated by the small percentage of terminally ill patients who choose PAS even where it is legal.

Opponents of legalized assisted suicide see it as an immoral slippery slope, worrying that it has grave potential for misuse and abuse. They feel it is dangerous for healthcare professionals, for our healthcare system, and for every patient. Among the dangers is the enhancement of distrust among patients and families who perceive themselves as disadvantaged and devalued by society. This concern leads to the fear that health systems will regard them as disposable and unworthy of care. Indeed, there is reason to believe this may be a real consequence of the legalization of PAS, based on our own country's history of abusing minorities (Dula 1994) and the experience of other societies that have permitted PAS and euthanasia (Lerner and Caplan 2015).

Public debates, especially in social media, have given the impression that large numbers of the public favor legalization of assisted suicide. Commenting on the pressure this has created, lawyer Wesley Smith has said:

> "If assisted suicide is now considered "courageous" and equates with "death with dignity," doesn't that imply that people … who choose to "fight against the dying of the light" are undignified and perhaps less courageous?[9]

Further support for this concern comes from Dr. Daniel Sulmasy, during an influential public debate on this topic. He quotes from Martin Luther King, whose grandmother told him, "Martin, don't let anybody ever tell you you're not a somebody." Sulmasy uses this theme to make the following claim: "Assisted suicide and euthanasia require us to accept that it is morally permissible to act with the specific intention-in-acting of making a somebody into a nobody, i.e., to make them dead." (Sulmasy et al. 2016) Based on his understanding, this violates the intrinsic dignity of us all (Fig. 10.3).

So what about the use of human dignity as an argument? It depends on the actual meaning of this ambiguous term. Meilaender has defined dignity in two ways. The first is "human dignity," simply defined as the characteristic of being human. On this view, human value can be quantified by function and is diminished by suffering and disease. The second variation is what Meilaender refers to as "personal dignity," which is tied into being an individual and is closely related to human equality. Such dignity is intrinsic and cannot be lost (Meilaender 2009). It is this latter sense that Sulmasy seems to have in mind in his argument against PAS.

Concepts of Dignity

Human Dignity:

- quantified by function
- diminished by suffering and disease

Personal dignity:

- intrinsic to all human beings
- cannot be diminished or lost

---

[9] Smith WJ. Of Michael Landon and Brittany Maynard. First Things Web site. www.firstthings.com/web-exclusives/2014/10/of-michael-landon-and-brittany-maynard. Published 2014. Accessed 12/19/2020.

Fig. 10.3 Dr. Daniel Sulmasy. (Personal communication, photo used with permission)

Many have made a utilitarian case against PAS as well. Dr. Ilora Finlay points out that determinations of prognosis at the end of life are notoriously inaccurate. Furthermore, 40% of the seriously ill have mental health issues, and dying patients often deal with depression. Even advocates for PAS will admit that mental health needs are often inadequately addressed. Under such circumstances, the existence of a PAS statute can create its own rationale. The coercion may be subtle, through high healthcare expenses at the end of life, the expiration of insurance, caregiver fatigue, and a sense of being a burden, not to mention unrecognized internal emotional issues.[10]

In their public debate on assisted suicide, Drs. Finlay and Sulmasy both make a passionate, slippery-slope case. They claim that PAS will inevitably lead to pervasive medical killing, with the endangering of vulnerable populations, including the disabled, the elderly, minorities, and the poor, i.e., anyone whose lives are seen as a burden on society.[9] Is there any actual data to reinforce this concern?

Lerner and Caplan reviewed reports from an end-of-life clinic in the Netherlands. Over a one-year study period, 645 patients applied for PAS or euthanasia and 162 were accepted. Though psychological suffering *per se* was rejected as a reason, 6.8% reported being "tired of living" as the sole reason for their request, and this was one of the reasons among 49.1% of the group. In 53.7% of the study population, approvals were in patients 80 years of age and older. The two authors, therefore, comment on the possibility of ageism influencing physicians approving the requests (Lerner and Caplan 2015). In their analysis of the Dutch data, Lerner and Caplan also point out that 1 in 30 deaths in the Netherlands was by euthanasia in 2012, three times the percentage from 2002, with the definition of "unbearable suffering"

---

[10] Debate on Assisted Suicide. Intelligence Squared Web site. www.intelligencesquaredus.org/debates/legalize-assisted-suicide. Published 2014. Accessed 12/19/2020.

expanding every year (Lerner and Caplan 2015). These facts lend support to the idea of a slippery slope.

In the same article, the authors turn to statistics from Belgium, once again with disturbing results. In Belgium, 1.9% of all deaths were from euthanasia in 2007. By 2013, the rate was 4.6% or 1 in 22 deaths (higher than in the Netherlands). The approval rate of requests jumped from 55% to 77% in the same period. "Tiredness of life" was allowed as a reason in 2013, though not in 2007. The fastest-growing numbers were among women, the elderly, those with less education, and nursing home residents. Note that this analysis points out one of the problems with slippery-slope arguments – they are most convincing when made retrospectively. In the words of Lerner and Caplan, "Part of the problem with the slippery slope is you never know when you are on it." (Lerner and Caplan 2015) Slippery-slope concerns against PAS are perhaps best summed up by Wesley Smith: "Will the 'right to die' evolve into a 'duty to die?'"[11]

Ethical Arguments in Favor of Assisted Suicide

In contrast to the abstract principlism of the opponents of PAS, those who favor PAS tend towards a more pragmatic approach. An excellent example of these arguments comes from Andrew Solomon.[9] He attacks the Hippocratic tradition, which he claims has no real philosophical underpinnings beyond mere tradition. There are religious overtones, to be sure, but these come from an outmoded and discredited Greek pantheon of deities. Furthermore, we are now a mostly secular society.

According to Solomon, there is a heavy-handedness to the Hippocratic Oath, which is paternalistic and elitist. The absence of autonomy is in stark contrast to our current focus on patient-centered care. Though many claim that the Oath has a pro-life emphasis, this is debatable and certainly does not represent the present diversity of ethical opinions in the medical profession. The modern health care environment is, in fact, much more complex than the ancient context.

Solomon finds appeals to human dignity unconvincing. It would appear that he sees dignity as a functional quality that can be diminished by suffering and disease (Meilaender's "human dignity"), as opposed to Sulmasy's claim that dignity is intrinsic (Meilaender's "personal dignity") (Meilaender 2009). On Solomon's understanding, not everyone can find meaning in suffering, and physical deterioration may lead to a loss of personal meaning.[9]

It is, therefore, understandable that patients may want to exert more control over their final moments. In Solomon's view, hospice and palliative care specialists are arrogant and authoritarian in their denial of the last wishes of dying patients. Furthermore, it would seem that about 70% of the U.S. and Canadian population

---

[11] Smith WJ. Lethal Ageism. First Things Web site. www.firstthings.com/web-exclusives/2014/10/lethal-ageism. Published 2014. Accessed 12/19/2020.

Fig. 10.4 Dr. Peter Singer. (By Crawford Forum – IMG_4421_Peter_Singer, CC BY 2.0, https://commons.wikimedia.org/w/index.php?curid=68668469)

agree with him, based on several recent opinion polls.[12, 13, 14] Solomon concludes his passionate appeal for the legalization of PAS by saying, "You want to support my dignity? Then respect my choices."[9]

Philosopher Peter Singer, for his part in the assisted suicide debate, looks squarely at the issue of death itself. Death violates our autonomy and is, therefore, bad for us and for those who care about us. Consequently, we should not promote death in medical practice. But, for Singer, assisted suicide is an exception to all this. Death may reasonably become our autonomous choice when there is no more value to continued life, i.e., when we cannot look forward to more life (Fig. 10.4).[9]

According to Singer, hastening death is not at the heart of the ethical concerns over PAS. Consider three types of intervention at the end of life: (1) withdrawal of futile treatment, (2) terminal (palliative) sedation, and (3) medical aid in dying (PAS or euthanasia). Each of these has the same final result. So what makes the first two legal and ethical? The key idea is *intent*. But intent is subjective, unmeasurable, and mostly irrelevant to the patient; the final result is the same. Why do we deny a comfortable death to those who do not fit the requirements for the first two types of intervention? Ethics professionals should find these arguments compelling.

---

[12] Campaign for Assisted Dying. www.dignityindying.org.uk/assisted-dying/public-opinion/. Published 2017. Accessed 12/19/2020.

[13] Euthanasia Still Acceptable to Solid Majority in U.S. Gallop Poll Web site. www.gallup.com/poll/193082/euthanasia-acceptable-solid-majority.aspx. Published 2016. Accessed 12/19/2020.

[14] Most Canadians Believe a Doctor Should be Able to Assist Someone Who is Terminally Ill and Suffering Unbearably to End their Life. Ipsos Web site. https://www.ipsos.com/en-ca/most-84-canadians-believe-doctor-should-be-able-assist-someone-who-terminally-ill-and-suffering. Published 2014. Accessed 12/19/2020.

On Singer's point of view, integral to the highest appeal to human liberty should be the right to die as one chooses and the right to decide when life is no longer worth living. This stance is simply a matter of personal autonomy. Neither government nor religious institutions should impose their values on others, especially on those who are not causing harm. As an option in end-of-life care, PAS would allow terminally ill, mentally competent individuals to retain dignity and bodily integrity in the face of unbearable pain and suffering.[9]

But does the exercising of such rights come at too high a cost? Is there actually a slippery slope? Let us assume, for the sake of argument, that data from Belgium and Netherlands truly show some worrisome trends. Claiming that the same thing will happen here in the U.S. is a logical fallacy, that of *conflation*, i.e., comparing apples to oranges. The small European countries that have legalized PAS are much more socially homogenous than the U.S. There are greater controls in our country, and all of our existing statutes have safeguards designed to prevent abuse.[9]

For example, consider Oregon, the state with the most experience with legal PAS. As of 2014, 859 patients died as a result of assisted suicide in the state, according to a recent comprehensive review. Early on, 1 in 1000 deaths in Oregon was from PAS. By 2014 it was 3 in 1000. In 78%, the diagnosis was cancer; in 8%, it was ALS. Of those patients who took a lethal dose of medication, 90% were in hospice, and 95% died at home. One particular claim by the study authors seems well warranted: PAS does not undermine hospice and palliative care. Furthermore, rates of depression were *lower* in patients who requested PAS than in other hospice patients (Ganzini 2016).

Why might patients wish to exercise their autonomy in this way? Pain is not the most important reason to request PAS. Many patients just want more control over the trajectory of their final days. Where physician aid in dying is illegal, some may even hoard certain drugs (e.g., pain meds), then take an overdose early, fearing that they will lose the right as they become more incapacitated. If PAS becomes more generally available, patients will be offered more appropriate drugs at the right time. Of note, not all patients who are prescribed lethal prescriptions actually take them. For example, 2015 statistics from Oregon show that 218 patients received prescriptions, while only 132 (61%) subsequently ingested them and died.[15] Statistics from the same year in Washington State revealed that 213 patients received the drugs, while 166 (78%) took them.[16] In other words, it appears that peace of mind was an essential factor for some of the patients, even if they never actually took the drugs. Personal autonomy figures prominently in the case for physician-assisted suicide. In the face of unbearable suffering, proponents argue that PAS as a legal option should

---

[15] Oregon Health Authority Releases 2015 data summary on death with dignity act. Death with Dignity Web site. www.deathwithdignity.org/news/2016/02/2015-report/. Published 2016. Accessed 02/19/2020.

[16] Out Now: The 2015 Washington Death with Dignity Act Annual Report. Death with Dignity Web site. www.deathwithdignity.org/news/2016/08/2015-washington-annual-report/. Published 2016. Accessed 12/19/2020.

be available so that dying patients can retain their sense of dignity and control, deciding for themselves how their lives should end.

In summary, those who oppose PAS argue that it is an affront to intrinsic human dignity and violates long-established ethical traditions. According to Finlay, Sulmasy, and others, prescribed suicide erodes the doctor-patient relationship and potentially leads us down an immoral slippery slope. At least in their present forms, PAS laws are dangerous to vulnerable minorities and do not adequately address mental health concerns.

Those who support PAS argue that we should re-think how our society looks at death. On the viewpoint of Solomon, Singer, and others, opposing PAS is arrogant and fails to respect individual autonomy. A majority of Americans support PAS, and there is no empirical evidence in the U.S. for a slippery slope. There are adequate safeguards to protect the vulnerable, including addressing their mental health needs.

The main issue driving the debate over PAS in the U.S. context is an unfortunate trend towards futile overtreatment and a prolonging of the dying process. On this, both sides will agree. In response, the medical profession and society should provide all of the following:

- Better training in pain management.
- More enlightened narcotics laws for the seriously ill, to allow for adequate pain control.
- Better training for providers on how to treat and diagnose depression.
- Broader availability and increased funding of hospice and palliative medicine.
- More energetic promotion of careers in palliative care for physicians, nurses, and pharmacists.
- Improved training in end-of-life psychiatric, social, and pastoral care.

The elderly are becoming a higher percentage of our population, and medical technologies allow for longer life. How and when to permit death to occur has become a critical philosophical, theological, and practical debate. Whether or not medically-assisted death becomes a more significant part of our healthcare tradition, it is clear that the benefits of optimal medical care for the dying will be distributed inequitably. Therefore, some patients will benefit, and others will suffer.

Pharmacist Participation in Lethal Injection

We now turn to a fascinating yet disturbing modern dilemma that directly impacts pharmacy practice. You may recall the hypothetical case study we briefly presented in Chap. 1:

> Dr. Andreas Marangos is a pharmacist who chairs the Ethics Committee of a mid-sized community hospital. The state attorney general has asked him to approve a contract for compounding a three-drug protocol used in lethal injections for capital punishment. Three inmates are currently on death row in the state prison, with one execution scheduled for the following month. How should Dr. Marangos respond?

Our justice system is embroiled in controversy over a procedure that, at least until recently, a majority of Americans have supported: the death penalty. Older methods, including firing squad, electric chair, or gas chamber, have been supplanted by lethal injection. This trend has been thought to be more humane, and therefore less in conflict with the Eighth Amendment proscription against "cruel and unusual punishment." The actual amendment reads as follows: "Excessive bail shall not be required, nor excessive fines imposed, nor cruel and unusual punishments inflicted."[17]

But the use of lethal injection has brought with it a whole new set of issues, as revealed by several "botched executions" in recent years. In Ohio, for example, a January 2014 execution of rapist-murderer Dennis McGuire took 26 min and was called by one defense attorney, "a failed, agonizing experiment."[18] In April of that same year, Clayton Lockett's lethal injection procedure by Oklahoma officials lasted 43 min and only ended when the condemned suffered a massive heart attack (Fig. 10.5)[19].

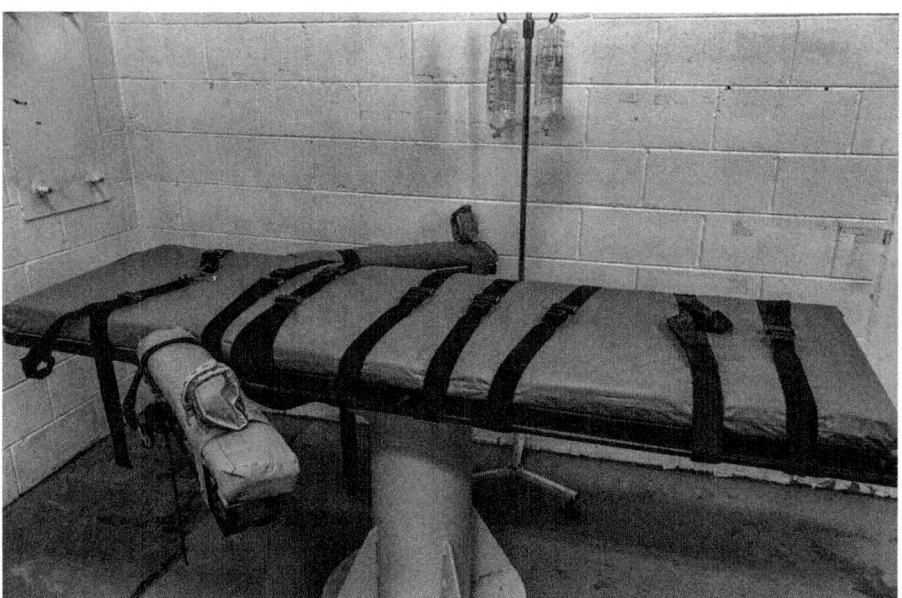

Fig. 10.5 Lethal injection table. (Credit: Ken Piorkowski – Lethal injection table. CC BY-SA 2.0.)

---

[17] Stevenson BA, Stinneford JF. The Eighth Amendment. Interactive Constitution. 2020. https://constitutioncenter.org/interactive-constitution/interpretation/amendment-viii/clauses/103

[18] Strauss G. Ohio killer's slow execution raises controversy. USA Today. www.usatoday.com/story/news/nation/2014/01/16/ohio-killer-executed-with-new-lethal-drug-combo/4512651/. Published 1/14/2014.

[19] Muskal M. Private autopsy blames Oklahoma for botched execution. LA Times. www.latimes.com/nation/nationnow/la-na-nn-autopsy-botched-oklahoma-execution-20140613-story.html. Published 6/14/2014.

Most Americans have historically supported the death penalty, but that has dramatically changed in recent years. According to Gallup polling, 61% of Americans were in favor of the death penalty in 1998. By 2019, support had dropped to 36%, the lowest in recent history. This trend is a massive shift in public opinion. For the healthcare profession, two other significant recent changes are worth noting.

The first is a renewed call for clinical professionals to get involved. After all, correctly applied professional medical judgment could ensure that the condemned prisoner is truly unconscious before injecting agents to stop the heart or to suppress respirations. This approach would significantly reduce fears of violating Eight Amendment safeguards. A second major shift is the shortage of drugs used for lethal injection. This change has happened because many European and American companies have blocked the use of their products for executions. In May of 2016, Pfizer joined the list, closing the last open-market source of drugs for possible use in executions (Dyer 2016).

To mitigate the issues associated with the McGuire case, Ohio amended its lethal-injection protocol. Still, capital punishment remains on hold, and momentum is building for its repeal in that state.[20] As of 2020, capital punishment remains legal in 29 states, though a number have governor-imposed moratoria.[21] Some states have reached out to compounding pharmacies to produce the agents, to shield from manufacturers the fact that their drugs were used in executions (Dresser 2014).[22] These trends have created significant legal and ethical controversy.

Though the ethics of capital punishment itself is beyond the purpose of this chapter, it is rightfully within our purview to consider whether pharmacists should play a role. In one sense, this seems very appropriate, as pharmacists are experts in medication; of all clinicians, they are the ones most qualified to select proper drugs, along with the correct dose and route of administration for each one. Yet there is a problem here, for the chemical agents used in a lethal injection cocktail cannot legally or ethically be called *drugs* since they have no clear beneficent purpose. According to Title 21 of the Food, Drug, and Cosmetic Act (1938, as amended), the term *drug* means, among other things, an "[article] intended for use in the diagnosis, cure, mitigation, treatment, or prevention of disease in man or other animals…"[23] None of these purposes is in view with the agents used in lethal injection for capital punishment.

Furthermore, one cannot refer to the condemned as a *patient* to whom professional obligations might apply. Indeed, there exists no covenantal or fiduciary duty

---

[20] Schladen M. In Ohio, talk of death penalty repeal is building. The Columbus Dispatch. www.dispatch.com/news/20200217/in-ohio-talk-of-death-penalty-repeal-is-building. Published 2/17/2020.

[21] States with and without the death penalty – 2020. Death Penalty Information Center Web site. https://deathpenaltyinfo.org/state-and-federal-info/state-by-state. Accessed 11/17/2020.

[22] Kroll D. Why don't pharmacy groups Condemn The Profession's Role In Capital Punishment? *Forbes Online.* www.forbes.com/sites/davidkroll/2014/05/14/why-dont-pharmacy-groups-condemn-practitioner-role-in-lethal-injection/#379103691c29. Published 5/14/2014

[23] Food Drug and Cosmetic Act. 21 US Code § 321. 1938. www.law.cornell.edu/uscode/text/21/321

between a pharmacist and the recipient of a lethal injection. Therefore, a strong line of ethical and legal arguments would militate against any participation by Dr. Andreas Marangos in the furtherance of such a program.

At this point, it is worth noting that the American Medical Association's Code of Medical Ethics (Opinion 9.7.3) prohibits involvement by physicians in lethal injection: "[A]s a member of a profession dedicated to preserving life when there is hope of doing so, a physician must not participate in a legally authorized execution." The Code goes on to itemize a long list of specific professional activities ethically forbidden in this context.[24]

Pharmacy professional associations are less categorical in their stance. For example, APhA puts it this way: "The American Pharmacists Association discourages pharmacist participation in executions on the basis that such activities are fundamentally contrary to the role of pharmacists as providers of health care."[25] The International Academy of Compounding Pharmacies (IACP) has a similar statement:

> While the pharmacy profession recognizes an individual practitioner's right to determine whether to dispense a medication based upon his or her personal, ethical and religious beliefs, IACP discourages its members from participating in the preparation, dispensing, or distribution of compounded medications for use in legally authorized executions.[26]

Because the process of lethal injection is not a therapeutic encounter, and because there certainly is no "prescription" involved, the issues surrounding pharmacist involvement remain ethically and legally murky. In the realm of medical principlism, participation violates autonomy and non-maleficence. Yet many clinicians may feel an understandable and compassionate impulse to help a condemned prisoner undergo execution in a way that ensures his or her dignity and comfort. They may also wish to ensure that such disasters as the Dennis McGuire or Clayton Lockett cases will never occur again. One recent review summarized the possibilities this way:

> Several potential avenues for pharmacist participation in lethal injection exist, including drug acquisition, admixture, and delivery. Pharmacists might also be involved in providing expertise regarding the selection of individual agents that might be used in lethal injection, including aspects of dosing, compatibility, and effective delivery (Romanelli et al. 2008).

Nonetheless, the involvement of pharmacists in capital punishment remains inherently problematic. As a society, we have generally agreed that hanging, the firing squad, the electric chair, and the gas chamber all violate the Eighth Amendment's proscription of "cruel and unusual." However, it does not currently

---

[24] Code of Medical Ethics Opinion 9.7.3: Capital Punishment. American Medical Association. 2020. www.ama-assn.org/delivering-care/ethics/capital-punishment

[25] Spinnler M. APhA House of Delegates Adopts Policy Discouraging Pharmacist Participation in Execution. American Pharmacists Association Web site. www.pharmacist.com/press-release/apha-house-delegates-adopts-policy-discouraging-pharmacist-participation-execution. Published 2015. Accessed 11/17/2020.

[26] IACP Adopts Position on Compounding of Lethal Injection Drugs. Pharmacy Times. www.pharmacytimes.com/news/IACP-Adopts-Position-on-Compounding-of-Lethal-Injection-Drugs. Published 3/24/2015.

appear that the "medicalization" of executions is a viable alternative. At its core, lethal injection potentially makes pharmacists and other clinicians complicit in a non-medical, non-therapeutic, and inherently violent act. It is therefore hard to ethically justify participation by these professionals.

Conclusion

This chapter has deeply considered the ethics of two very controversial areas of modern healthcare, both with significant implications for pharmacy practice. We began with a no-holds-barred examination of assisted suicide from both sides. We then moved on to the controversial question of whether pharmacists should be involved in lethal injection for capital punishment. The following chapter will examine the critical arena of healthcare rights of conscience.

Key Terms

- euthanasia
- assisted suicide
- slippery-slope argument
- human dignity v. personal dignity
- Eighth Amendment
- legal definition of a "drug"

Questions for Review

1. Senators in your state have proposed a "Death with Dignity" statute that would legalize the practice of assisted suicide.

    (a) A local APhA chapter has asked you to testify in front of a hearing of the Senate Health Committee *on behalf* of this legislation. Speaking as a pharmacist, summarize the major arguments in your testimony.

    (b) Now suppose that the local APhA chapter has asked you to testify *against* this legislation. Speaking as a pharmacist, summarize the major arguments in your testimony.

2. Do you agree ethically with the APhA and IACP Codes that lethal injection is to be discouraged for pharmacists? Would you participate, and if so, to what extent?

For Further Reading

Arther Dyck. *Life's Worth: The Case Against Assisted Suicide.* Eerdmans, 2002. Dr. Dyck is Emeritus Professor of Population Ethics in the School of Public Health at Harvard University. This compact, easy-to-read treatment gives the main ethical arguments to counter the modern trend towards medically-assisted death.

Peter Singer. *Writings on an Ethical Life.* Ecco Press, 2000. Every serious student of clinical ethics should read this book by Dr. Singer, the Australian moral philosopher and Ira W. DeCamp Professor of Bioethics at Princeton University. Though this slim volume is not specifically about end-of-life issues, the reader will find within it a challenging, often infuriating, defense of his atheistic, utilitarian viewpoint. The text includes several chapters endorsing his approval of PAS and euthanasia.

References

Bulger RJ, Barbato AL. On the hippocratic sources of Western medical practice. Hast. Cent. Rep. 2000;30(4):S4–7.
Cassell EJ. The nature of suffering and the goals of medicine. 2nd ed. New York: Oxford University Press; 2004.
Cherny NI, Fallon M, Kaasa S, Portenoy RK, Currow D. Oxford textbook of palliative medicine. 5th ed. Oxford: Oxford University Press; 2015.
Devettere RJ. Practical decision making in health care ethics: cases, concepts, and the virtue of prudence. 4th ed. Washington, DC: Georgetown University Press; 2016.
Dresser R. Drugs and the death penalty. Hast. Cent. Rep. 2014;44(1):9–10.
Dula A. African American suspicion of the healthcare system is justified: what do we do about it? Camb. Q. Healthc. Ethics. 1994;3(3):347–57.
Dyer O. Pfizer blocks sales of its drugs for executions. BMJ (Clin. Res. Ed.). 2016;353:i2791.
Edelstein L. The Hippocratic oath, text, translation and interpretation. Baltimore: The Johns Hopkins press; 1943.
Fass J, Fass A. Physician-assisted suicide: Ongoing challenges for pharmacists. Am. J. Health-Syst. Pharm. 2011;68(9):846–9.
Ganzini L. Legalised physician-assisted death in oregon. Qld. Univ. Technol. Law Rev. 2016;16(1):76–83.
Hughes MT. The pharmacist and medical aid in dying. Am. J. Health-Syst. Pharm. 2017;74(16):1253–60.
Lerner BH, Caplan AL. Euthanasia in Belgium and the Netherlands: on a Slippery Slope? JAMA Intern. Med. 2015;175(10):1640–1.
Meilaender G. Neither beast nor god: the dignity of the human person. New York: Encounter Books; 2009.
Romanelli F, Whisman T, Fink JL. Issues surrounding lethal injection as a means of capital punishment. Pharmacotherapy. 2008;28(12):1429–36.
Sullivan DM, Taylor RM. The ethical landscape of assisted suicide: a balanced analysis. Ethics Med. 2018;34(1):49–57.
Sulmasy DP, Travaline JM, Mitchell LA, Ely EW. Non-faith-based arguments against physician-assisted suicide and euthanasia. Linacre Q. 2016;83(3):246–57.
Sulmasy DP, Finlay I, Fitzgerald F, Foley K, Payne R, Siegler M. Physician-assisted suicide: why neutrality by organized medicine is neither neutral nor appropriate. J. Gen. Intern. Med. 2018;33(8):1394–9.
Swetz KM, Thorsteindottir B, Feely MA, Parsi K. Balancing evidence-based medicine, justice in health care, and the technological imperative: a unique role for the palliative medicine clinician. J. Palliat. Med. 2012;15(4):390–1.
Van Way C. Physician advocacy: the AMA annual meeting 2019. Mo. Med. 2019;116(4):279–81.

Chapter 11
Rights of Conscience

So far, this text has focused on normative ethics principles on which pharmacists generally agree. These concepts form the basis of a widely-held consensus on professionalism. But what happens when an individual pharmacist is confronted with a prescription or treatment directive that runs counter to personal religious or moral values? Should the clinician simply go along with the order, defer to another pharmacist, or refuse to comply with the request? This question is complex, and it is the topic of this chapter on rights of conscience (Sullivan 2019).

Healthcare rights of conscience, also referred to as *conscientious objection in health care*, can be defined as "the refusal to perform a legal role or responsibility because of moral or other personal beliefs."(Berlinger 2008) Such a stance is often based on a religious perspective, though this is not always the case. To understand the context that informs this issue, it will be helpful to consider healthcare oaths once again.

You may recall that Chap. 7 reflected on professionalism, as exemplified in the Ethics Code of the American Pharmacists Association, strongly influenced by the ancient Hippocratic Oath. Let us take a moment to consider another modern oath taken by entering students at Washington University School of Medicine. It begins as follows:

> I will make my patient my foremost consideration as I strive to promote health and quality of life through education, prevention, and care;
>
> I will value the trust my patients place in me and endeavor to build that trust through compassion, honesty, and excellence;
>
> I will recognize, seek to understand, and respect each patient's individuality and autonomy;
>
> I will educate and empower my patients and their families to make choices that honor their values and beliefs;
>
> I will listen to my patients so that I may learn how to best serve them. (2017 White Coat Ceremony - Doctor of Medicine 2020)

The beautiful language of this declaration reflects idealism, gratitude, and humility, and is similar to oaths taken by future doctors across the United States. One recent survey revealed that more than half of American medical schools feature a customized oath, often personally written by the students, allowing them a more profound sense of engagement and ownership (Weiner 2018). However, there are some subtle dangers to this approach. Greiner and Kaldjian warn that "personal satisfaction gained by local authorship [should] not lessen the connection to the enduring values of the profession at large."(Greiner and Kaldjian 2018)

In other words, there may be something missing from such a modern oath. As we discussed back in Chap. 4, the original Hippocratic Oath begins this way:

> I swear by Apollo Physician, and Aesculapius, and Hygeia, and Panacea and all the gods and goddesses, making them my witnesses, that I will fulfil according to my ability and judgment, this oath and this covenant. (Edelstein 1943)

Notice the vertical dimension, in contrast to most modern oaths. Granted, most people are not familiar with the invoked deities that may seem rather quaint by today's standards (Apollo? Aesculapius?). And our modern pluralistic society will not readily agree on a common god or gods by which one should make such a pledge. Yet, at the very least, the words seem to carry greater weight than both the individual and the public at large (Greiner and Kaldjian 2018).

The language of the Hippocratic Oath implies binding duties that go beyond a mere promise. As we emphasized in Chap. 7, these duties entail a *fiduciary relationship*, one of trust, confidence, and loyalty. Out of the Oath spring the ancient obligations of *beneficence* (acting in the best interests of patients) and *non-maleficence* (avoidance of harm), respectively.(Kinsinger 2009; Gillon 1985) Furthermore, the Oath goes on to pledge commitments to justice, respect for one's teachers, personal integrity, and confidentiality, all central values that caused an obscure pagan creed to become the foundation for the normative practice of medicine for over two millennia (Pellegrino 2005).

The modern oath from Washington University mostly focuses on patient autonomy, and therefore seems to omit the "enduring values of the profession at large." While autonomy is undoubtedly significant, the ancient principles from the Hippocratic tradition also have considerable value and should be emphasized at least as much (Fig. 11.1).

Retaining ancient Hippocratic principles is not without its problems, however. If we blindly adhere to the Oath or even use it as a model for pharmacy practice, there may be several shortcomings. Here are a few:

1. The School of Hippocrates only allowed men to become practitioners, making it potentially paternalistic and elitist.
2. The Oath is vague, rigid, and authoritarian.
3. We now live in a more secular society, uncomfortable with covenantal language.
4. The modern approach to health care is patient-centered, whereas Hippocratic medicine revolved around the physician's personal integrity.
5. The practice of medicine is now much more complex than its ancient underpinnings, and its ethics should account for these changes.

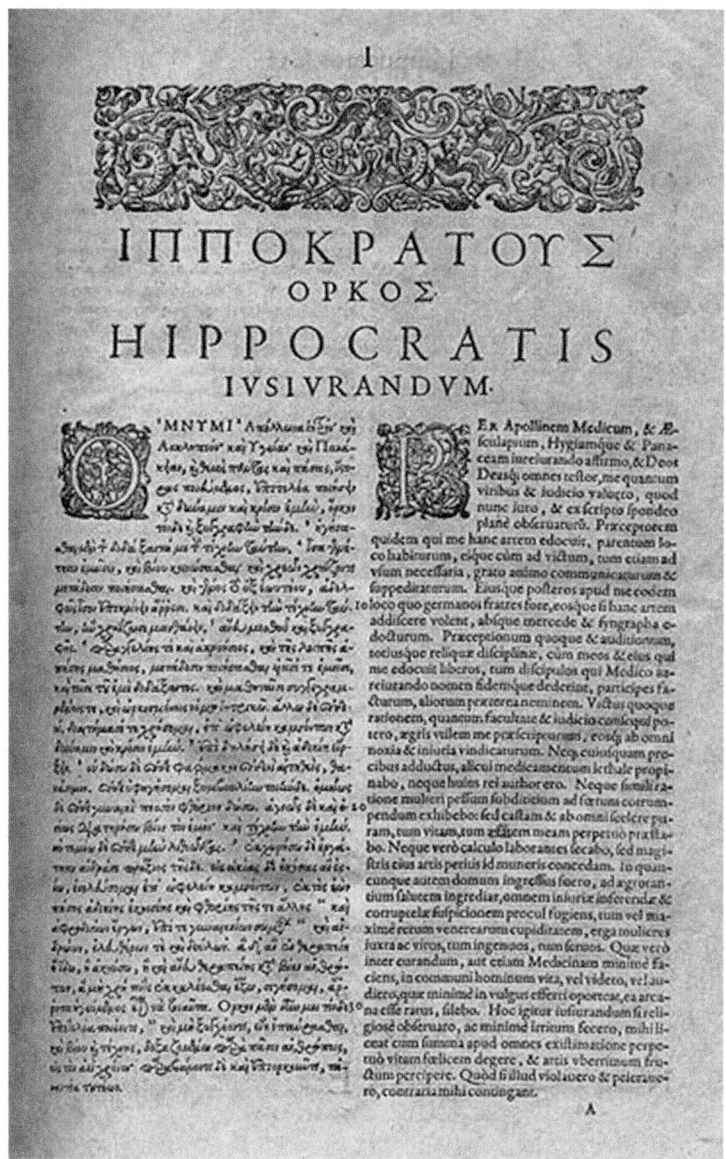

Fig. 11.1 The Hippocratic Oath in Greek. (Public Domain: the Hippocratic Oath in Greek and Latin published in Frankfurt in 1595 in Apud Andreae Wecheli heredes by Claudium Marnium, & Ioan. Aubrium)

The modern perspective on professionalism in medicine, and by extension in pharmacy practice, therefore, draws on the rich heritage of Hippocrates but also adds the more recent egalitarian principles of autonomy, informed consent, and patient-centered care. This approach, as we have seen, is now well-known as *medical principlism*, and includes four tenets, combining old and new: beneficence, non-maleficence, distributive justice, and autonomy (Beauchamp and Childress 2019).

To make this work in today's practice environment requires a commitment to normative ethics. Though we come from a wide variety of philosophical and religious backgrounds, there is a clear-cut mechanism for navigating tough ethical dilemmas (Fieser 2018). This approach places high trust in the clinician's moral judgment. Indeed, this is the basis of professional *virtue*, discussed in Chap. 7, which focuses on the personal character of individual pharmacists and the wisdom that such character entails (Kraut 2018).

And so we come to the topic of this chapter: rights of conscience. Healthcare professionals have become increasingly concerned about protections for religious freedoms in the workplace. A new federal government office called the Conscience and Religious Freedom Division is situated under Health and Human Services. The Office for Civil Rights enforces its standards (Conscience Protections for Health Care Providers 2018). New rules protect health care workers (physicians, nurses, pharmacists, etc.) who hold to religious or moral objections to certain medications, treatments, or procedures. These rules allow them to opt-out of providing such services based on conscience.

But how should we define healthcare conscience claims? Patient autonomy is undoubtedly essential in medicine, but is there such a thing as professional autonomy for clinicians? In general, rights of conscience are couched as "negative rights," i.e., the right to refuse to provide a requested treatment. Indeed, there are many well-established professional reasons for this stance. Robert Orr has explained it this way:

> [Clinicians may] refuse on the basis of a legal standard (e.g., a request for a non-approved drug), a professional standard (e.g., a request for non-beneficial treatment such as hyperbaric oxygen for a completed stroke), clinical judgment (e.g., a request for an antibiotic to treat a viral infection), or even a personal choice (e.g., in non-emergency situations, doctors are free to refuse patients for nondiscriminatory reasons). (Orr 2013)

Outside of these circumstances, do clinicians have a further right of refusal? An ethics opinion from the American Medical Association certainly implies that they do:

> Preserving opportunity for physicians to act (or to refrain from acting) in accordance with the dictates of conscience in their professional practice is important for preserving the integrity of the medical profession as well as the integrity of the individual physician, on which patients and the public rely. Thus physicians should have considerable latitude to practice in accord with well-considered, deeply held beliefs that are central to their self-identities. (Physician Exercise of Conscience 2019)

What about pharmacists? In the light of controversies over contraceptive and abortifacient drugs, the American Pharmacists Association adopted a policy to "ensure patient access to legally prescribed therapy without compromising the pharmacist's right of conscientious refusal."(Buerki et al. 2013) The American Society of Health-System Pharmacists, in its Council on Legal and Public Affairs, explains it as follows:

[We] recognize the right of pharmacists, as health care providers, and other pharmacy employees to decline to participate in therapies they consider to be morally, religiously, or ethically troubling…(American Society of Health-System Pharmacists 2020)

These ethical standards are well-established but are not necessarily backed up by state and federal laws, especially concerning pharmacy practice. One legal scholar points out the wide variability of state laws, some endorsing, and some limiting conscience claims. He adds this warning:

Given the varying regulatory approaches by the states on the provision of pharmacy care, at least with respect to contraception, pharmacists should review the applicable pharmacy-practice acts and other statutes and regulations for guidance. A pharmacist who is considering whether to refuse services for ethical, moral, or religious reasons should carefully review those laws, regulations, and administrative and court rulings. Depending on the state where the pharmacist practices, a refusal to provide services on those grounds may result in judicial and regulatory scrutiny. (Langjahr 2018)

So with the caveat that well-thought-out ethical standards may not be supported by state or federal statutes, let us analyze the ethical principles that may undergird conscience claims. We must point out, as we have often done in this text, that ethics requires the balancing of competing duties. In this era of heightened patient autonomy and choice, there is a great deal of pressure on clinicians to cave to all patient desires. By way of pushback, doctors, nurses, pharmacists, and other healthcare professionals would claim they are not mere "vending machines" for medications and procedures, but caring individuals who have a stake in their patients' lives. Rights of conscience are an expression of accommodations for professionals to decline to perform, assist with, or provide specific objectionable treatments or procedures.

However, such accommodations are not unlimited. Conscience rights should be respected and protected by law, but patients have a right to legal treatments and may not be abandoned by the healthcare system. The debate is further complicated by the fact that it centers on highly controversial matters: abortion, sterilization, contraception, assisted suicide, gender-transitioning drugs, and many others. A key question is simply this: are some patients denied care, or are they simply inconvenienced by the refusal of some clinicians to provide these services?

To answer this question, let us first consider an example from obstetrical medical practice. Ethics Committee Opinion #385 of the American College of Obstetricians and Gynecologists (ACOG) states that obstetricians are morally obligated to refer patients for abortions. If another clinician is not available, then a physician must go ahead and perform the procedure, regardless of one's personal views (ACOG Committee 2007). Philosopher Robert George responds to this imperative by challenging the nature of elective abortion itself as routine healthcare, based on the implicit judgment that pregnancy, when unwanted, is, in effect, a disease. The issues in dispute are philosophical; science or methods of scientific inquiry cannot resolve them. In the words of Dr. George: "If this is the case, there is, therefore, no justification to compel morally sincere physicians who disagree with this judgment to violate their conscience or else leave the practice of medicine entirely," as some moral philosophers and lawmakers have suggested. His conclusion is succinct: "The

ACOG statement fails to acknowledge the widespread debate about abortion in our society and the moral sincerity of pro-life practitioners."(George 2019) In this way, Dr. George acknowledges the legal right of women to an abortion, while affirming the moral right of clinicians not to participate.

We now turn to a challenging case from pharmacy practice. Every clinical encounter involves at least two moral agents: the clinician and the patient. A conflict may arise when the two agents are ethically at cross-purposes. In the summer of 2018, Nicole Arteaga, a pregnant Arizona patient, was diagnosed with fetal demise *in utero* during her first trimester. The patient's physician prescribed misoprostol, a prostaglandin used to induce uterine contractions. The clinical goal was to cause the delivery of the dead fetus, thereby eliminating the need for a surgical procedure. She presented the prescription to the pharmacist at Walgreens in Peoria, Arizona, and that is where things went amiss. The Walgreens pharmacist questioned Arteaga extensively as to the reason for the prescription, and the questioning took place in front of her 7-year old child and the other customers waiting in line. His actions resulted in embarrassment and humiliation for Arteaga. The pharmacist then refused to fill the prescription because of his personal beliefs. His refusal persisted, despite explanations by both the patient and her husband. His specific reasons remain unknown. The patient filled the prescription the next day at a different Walgreens (Lucero 2018; Chuck 2018).

Like several other states, Arizona has a law protecting rights of conscience that may cause some healthcare providers to opt-out of dispensing abortifacient medications. Such beliefs must be laid out in writing ahead of time with the employer (NCSL 2018). In short, the pharmacist had a legal right to decline to fill the prescription, had he handled the situation correctly. What went wrong in this particular case?

Pharmacists are well-trained in history taking, and this is necessary to ensure the safety of patients and the efficacy of therapy. Getting more information is entirely in keeping with best practices. It is in the patient's best interests that a pharmacist should do so, and of course, any information must be kept in strict confidence. In this case, however, it appears that the pharmacist had already made up his mind to refuse to dispense misoprostol when he began to question the patient. This is unfortunate because a careful history might have demonstrated to him that fetal demise had already occurred, so dispensing the drug would not have violated his beliefs. Most religious belief systems allow medications for miscarriage and treatment when the life of the mother is in danger. Though not an issue in this case, these belief systems usually permit such treatments even when the life of the fetus is also at risk (Ethical and Religious Directives for Catholic Health Care Services 2018).

The ethical question here is a matter of competing interests. On whose interests was the pharmacist acting? If acting on the interests of his patient, then further questions to clarify the clinical intent of the medication were appropriate. This approach would meet the goal of proper counseling on the risks and safe use of the drug. However, this role should be reserved to the pharmacist who actually fills the prescription. By making an *a priori* decision not to dispense, the pharmacist relinquished this role and began to act primarily on his own interests, violating his ethical duty.

There is a further ethical lapse in this matter. In handling sensitive, protected patient information, pharmacists are bound ethically to ensure patient confidentiality. This duty is further laid out as a legal requirement by the Health Insurance Portability and Accountability Act (HIPAA) (History of HIPAA 2018). While some routine exchange of information typically happens at the pharmacy counter, most contemporary pharmacies provide consultation areas where more extensive questioning may occur. Sensitive information may be exchanged in those places to preserve confidentiality. Assessing the patient in front of a child and within earshot of other customers was a violation of confidentiality and non-maleficence. Ethically, the pharmacist handled the situation poorly; legally, it was a violation of HIPAA.

By way of contrast, the case for a conscience claim by a pharmacist is a bit more apparent in the case of an elective first-trimester medical abortion. As we noted in Chap. 8, a medical abortion can be performed up to 10 weeks of gestation, using two medications: mifepristone, a progesterone-receptor antagonist, usually given to the patient by the gynecologist, (Justine 2010) and misoprostol, a prostaglandin dispensed by a community pharmacist (Planned Parenthood Web site 2020). If a pharmacist has a moral objection to filling the misoprostol prescription, he or she should politely refer the patient to another pharmacist, hopefully at the same location. The goal should be to minimize patient inconvenience or embarrassment (Anderson et al. 2006).

Similar conscience concerns have arisen concerning contraceptives, especially emergency contraceptives, since some clinicians believe that they may be abortifacient, at least some of the time. In addition, it is also abundantly clear that many clinicians disagree with medical aid in dying (i.e., physician-assisted suicide), and some pharmacists may wish to decline to dispense the drugs that make this possible, such as a lethal dose of secobarbital.

It is generally accepted that referral of the problematic prescription to another on-duty pharmacist is the best way to handle such problems with personal conscience. However, the perceived moral wrong of a particular treatment may so offend the conscience of a clinician that he or she wants no part in it, raising the issue of moral *complicity*. Complicity refers to the possible taint of moral guilt attached to a person by association with a moral wrong. For example, an accomplice or accessory to a crime is just as guilty in the eyes of the law as the person who performs the deed (Accessory after the fact 2020). From a moral perspective, complicity implies that a person has some association with a morally wrong act, even if he or she did not personally perform that act. Such a situation might incline a pharmacist to be unable to refer to another pharmacist, instead preferring to return the prescription to the patient. Because this approach seems to create a higher barrier for patients to receive legally-available medications, it has not been widely accepted, and it is still hotly debated among clinicians and ethicists (Brock 2008; Hughes 2018; Riker 2015).

One final point is worth making about the shared burdens of conscience. If pharmacists must be flexible and careful in the way they make conscience claims, prescribers must be as well. Physician Kevin Powell points out that "reasonable accommodation is a characteristic of the law in a morally just society." He goes on to make this insightful and helpful observation:

> It is not an unreasonable burden to expect an obstetrician prescribing controversial treatments like abortifacients to know if a local pharmacist has a conscientious objection to providing those medications. Such knowledge by the obstetrician would prevent most cases of embarrassment and indignity when a female patient attempts to fill such a prescription. Similarly, we may reasonably expect a palliative care physician who prescribes a lethal dose of pills to be used for assisted suicide to know which nearby pharmacists are willing to fill that prescription. (Powell 2019)

So what does a virtuous clinician look like in today's healthcare environment? Are the ideals of Hippocrates completely outmoded, or merely in need of an update? Clinicians of deep conviction and conscience are sorely needed, but will they be driven away from their calling by the pressures of some understandings of autonomy and managed care?

We would call on pharmacists to remember their professionalism. The word 'profession' comes from the Latin *professio*, meaning a public oath of fealty, or turning over one's obedience and loyalty to another (Profession 2019). Pharmacists are not merely mechanics doing a job, for they have covenantal, binding obligations to their patients that call them to a high standard. If healthcare rights of conscience are to make sense, they must reflect this deep concern for the well-being of the patients under our care.

In the next chapter, we will examine an array of additional ethics questions, with particular emphasis on pharmacy practice.

Key Terms

- rights of conscience
- fiduciary relationship
- negative rights
- prescription referral
- HIPAA
- moral complicity
- profession

Questions for Review

1. What professional characteristics are implied by the argument that pharmacists are more than mere "vending machines?"
2. What does it mean to say that conscientious practice implies certain "negative rights?" Could a right of conscience ever be described as a "positive right?"
3. Define moral complicity. How might this concept affect your willingness to provide a specific clinical treatment?

For Further Reading

Giles Birchley. A Clear Case for Conscience in Healthcare Practice. *Journal of Medical Ethics.* 2012;38(1):13–17. The author invokes the writings of philosopher and political theorist Hannah Arendt to argue persuasively that conscience rights are not only permissible, but that they are desirable for the overall good of the healthcare profession. (Birchley 2012)

Marc Clauson. The emergence of conscience rights for health care professionals: How we moved historically from conscience to conscience rights. *Ethics, Medicine, and Public Health.* 2019;11:21–29. An extensive review of conscience rights by a historian and political philosopher. The author demonstrates how the idea of conscience began in ancient philosophy and Christian thought, associated with private and individual internal moral guidance. (Clauson 2019)

References

2017 White Coat Ceremony - Doctor of Medicine. https://whitecoat.wustl.edu/wcc2017. Published 2020. Accessed 14 Oct 2020.

Accessory after the fact. The Legal Information Institute (Cornell Law School). 2020. www.law.cornell.edu/uscode/text/18/3.

ACOG Committee Opinion No. 385 November 2007: The limits of conscientious refusal in reproductive medicine. Obstetrics and Gynecol. 2007;110(5):1203.

Pharmacist's Right of Conscience and Patient's Right of Access to Therapy. American Society of Health-System Pharmacists. 2020. www.ashp.org/Pharmacy-Practice/Policy-Positions-and-Guidelines/Browse-by-Topic/Ethics.

Anderson RM, Bishop LJ, Darragh M, Gray HH, Poland SC. Pharmacists and conscientious objection. Kennedy Inst Ethics J. 2006;16(4):379–96.

Beauchamp TL, Childress JF. Principles of biomedical ethics. 8th ed. New York: Oxford University Press; 2019.

Berlinger N. Conscience clauses, health care providers, and parents. From birth to death and bench to clinic: the hastings center bioethics briefing book for journalists, policymakers, and campaigns. Garrison: Hastings Center; 2008. p. 35–40.

Birchley G. A clear case for conscience in healthcare practice. J Med Ethics. 2012;38(1):13–7.

Brock DW. Conscientious refusal by physicians and pharmacists: Who is obligated to do what, and why? Theoret Med Bioethics. 2008;29(3):187–200.

Buerki RA, Vottero LD, American pharmacists association. Pharmacy ethics: a foundation for professional practice. American Pharmacists Association: Washington, D.C; 2013.

Chuck E. Walgreens pharmacist refuses to give Arizona woman drug to end pregnancy. NBC News. June 25, 2018, 2018. www.nbcnews.com/news/usnews/walgreens-pharmacist-refuses-give-arizona-woman-drug-end-pregnancy-n886396.

Clauson M. The emergence of conscience rights for health care professionals: How we moved historically from conscience to conscience rights. Ethics Med Pub Health. 2019;11:21–9.

Conscience Protections for Health Care Providers. 2018. www.hhs.gov/conscience/conscience-protections/index.html.

Ethical and Religious Directives for Catholic Health Care Services. United States Conference of Catholic Bishops, 2018. www.usccb.org/about/doctrine/ethical-and-religious-directives/upload/ethical-religious-directives-catholic-health-service-sixth-edition-2016-06.pdf.

Fieser J. Ethics. Internet encyclopedia of philosophy. 2018. www.iep.utm.edu/ethics/.

George R. A defense of conscience in healthcare. Ethics Med Pub Health. 2019;11:16–20.

Gillon R. "Primum non nocere" and the principle of non-maleficence. Br Med J (Clin Res Ed). 1985;291(6488):130–1.

Greiner AM, Kaldjian LC. Rethinking medical oaths using the physician charter and ethical virtues. Medical Education. 2018;52(8):826–37.

History of HIPAA. The HIPAA Guide Web site. https://www.hipaaguide.net/history-of-hipaa/. Published 2018. Accessed 10/11/2020.

How does the abortion pill work? Planned Parenthood Web site. www.plannedparenthood.org/learn/abortion/the-abortion-pill/how-does-the-abortion-pill-work. Published 2020. Accessed 31 Oct 2020.

Hughes JA. Conscientious objection, professional duty and compromise: A response to Savulescu and Schuklenk. Bioethics. 2018;32(2):126–31.

Justine W. Medication abortion using mifepristone and misoprostol. New York: Springer; 2010. p. 347.

Kinsinger FS. Beneficence and the professional's moral imperative. J Chiropractic Humanities. 2009;16(1):44–6.

Kraut R. Aristotle's Ethics. Stanford Encyclopedia of Philosophy 2018. https://plato.stanford.edu/entries/aristotle-ethics/.

Langjahr A. Women's reproductive health and the right to not dispense. US Pharm. 2018;43(9):34–5.

Lucero L. Walgreens pharmacist denies woman with unviable pregnancy the medication needed to end it. The New York Times. June 25 2018. 2018.

NCSL. Pharmacist Conscience Clauses: Laws and Information. (September, 2018). www.ncsl.org/default.aspx?tabid=14380.

Orr RD. Autonomy, conscience, and professional obligation. Virtual Mentor. 2013;15(3):244–8.

Pellegrino E. Some things ought never be done: moral absolutes in clinical ethics. Theoret Med Bioeth. 2005;26(6):469.

Physician Exercise of Conscience: Code of Medical Ethics Opinion 1.1.7 American Medical Association. 2019. www.ama-assn.org/delivering-care/ethics/physician-exercise-conscience.

Powell K. Reasonable accommodation of conscientious objection in health care is morally and legally required. Perspect Biol Med. 2019;62(3):489–502.

Profession. Miriam-Webster Online. 2019. www.merriam-webster.com/dictionary/profession.

Riker WJ. The complicity objection and the return of prescriptions. Southwest Philos Rev. 2015;31(1):207–16.

Sullivan D. Portions of this chapter first appeared in: professionalism, autonomy, and the right of conscience: a call for balance. Ethics Med Pub Health. 2019;11:11–5. Used with permission

Translation from the Greek by Ludwig Edelstein: The Hippocratic oath, text, translation and interpretation. Baltimore: The Johns Hopkins press; 1943.

Weiner S. The solemn truth about medical oaths. American Association of Medical Colleges. www.aamc.org/news-insights/solemn-truth-about-medical-oaths. Published July 10, 2018.

Chapter 12
Vaccines, Resource Allocation, and Unproven Treatments

So far, this text has examined the foundations of ethical practice in history and professionalism. We then considered several specific topics: reproductive ethics, end of life, and conscience rights. This chapter will focus on certain specific additional topics not previously discussed. Let's begin with the fascinating subject of vaccines and their ethical controversies.

The Ethics of Vaccines

Most of us who have studied biology have heard the story of Edward Jenner, a "simple country doctor from Gloucestershire," (Smith 2011) who, in the late eighteenth century, made a remarkable observation that led to the first vaccine. From the website for Dr. Jenner's House, a museum in Gloucestershire, UK, we read this somewhat adulatory summary:

> Born in Berkeley in 1749, Edward Jenner spent his life making remarkable discoveries, driven by an intense curiosity about how everything worked and a desire to make the world a better place. In 1798 he published *An Inquiry into the Causes and Effects of the Variolae Vaccinae, a Disease discovered in some of the Western Counties of England, particularly Gloucestershire, and known by the name of the Cow Pox*, making known his investigations into how a mild disease, cowpox, could protect against the horrific smallpox virus. From the Latin word *vacca*, meaning "cow," he called this new practice **vaccination**. (History 2020)

Jenner was a meticulous scientist, who made the remarkable observation that dairymaids previously infected with cowpox were resistant to smallpox. He put this hypothesis to the test with his famous experiment, which involved "taking pustular material from a cowpox lesion in a human infected from a cow, and inoculating a young boy." (The Eradication of Smallpox 2001) The boy, an eight-year-old named James Phipps, became resistant to smallpox, laying the foundation for the first truly

---

This chapter features an additional author, PharmD candidate Joshua Pearson (class of 2021).

safe and successful vaccine. Ironically, such an experiment would never be possible today, as Jenner had no prior consent from the boy's parents, and there was no ethical oversight (Chiswick 1996).

Today, the number of infectious diseases preventable by vaccines is vast. The CDC lists the following: diphtheria, *Haemophilus influenzae* type b, hepatitis A, hepatitis B, herpes zoster, human papillomavirus, influenza, measles, meningococcal infections, mumps, pertussis, pneumococcal infections, poliomyelitis, rotavirus, rubella, tetanus, and varicella, to name just a few (Vaccines & Immunizations 2018). To understand better the significant effect such vaccines have had, consider the overall reduction in morbidity from many of these illnesses, as shown in the following table (adapted from public health data Hinman 1999) (Table 12.1):

Despite these successes, public misinformation has often led to a significant decrease in vaccination rates, based on several fears. Consider this example from Japan:

> In 1974, about 80% of Japanese children were getting pertussis (whooping cough) vaccine. That year there were only 393 cases of whooping cough in the entire country, and not a single pertussis-related death. Then immunization rates began to drop, until only about 10% of children were being vaccinated. In 1979, more than 13,000 people got whooping cough and 41 died. When routine vaccination was resumed, the disease numbers dropped again. (Vaccines & Immunizations 2018)

Though the problems in Japan were multifactorial, misperceptions about side effects and efficacy of vaccines contributed to the reluctance of parents to vaccinate their children. Public awareness campaigns played a significant role in improving the situation (Kuwabara and Ching 2014). Nonetheless, false information about the safety of vaccines has severely hampered the widespread acceptance of immunization in many countries. Amazingly, most of these more recent issues can be traced back to a single researcher in England in 1998.

Andrew Wakefield, a former British gastroenterologist, was the lead author of a report in the medical journal *The Lancet*, published in 1998, that claimed a possible causal connection between the common vaccine against measles, mumps, and rubella (MMR) and childhood autism (Wakefield et al. 1998). Though based on only 12 children, Wakefield made strong claims about the dangers of the MMR vaccine, even announcing these ideas in a press conference before publication of the *Lancet* report (Triggle 2010).

Table 12.1 Maximum and current morbidity of selected diseases

Disease	Maximum	1997	Decrease
Diphtheria	206,939	4	99.99%
Measles	894,134	138	99.98%
Mumps	152,209	683	99.55%
Pertussis	265,269	6564	97.52%
Polio (paralytic)	21,269	0	100.00%
Rubella	57,686	181	99.69%
Congenital Rubella (est.)	20,000	5	99.68%
Tetanus (mortality)	1560	50	99.79%
H. influenzae (peds., est.)	20,000	165	99.18%

The outcome for the public in the UK was devastating. The resulting vaccine scare led to a significant decline in public confidence, and vaccine rates dropped precipitously. Cases of measles, previously rare, now soared. From the *British Medical Journal*:

> In 2008, for the first time in 14 years, measles was declared endemic in England and Wales. Hundreds of thousands of children in the UK are currently unprotected as a result of the scare, and the battle to restore parents' trust in the vaccine is ongoing (Godlee et al. 2011).

In 2004, investigative journalist Brian Deer, writing for the *Sunday Times*, unearthed serious questions about Wakefield's original paper, finding it deliberately fraudulent and misleading (Deer 2004). Again from the *British Medical Journal*:

> Deer unearthed clear evidence of falsification. He found that not one of the 12 cases reported in the 1998 *Lancet* paper was free of misrepresentation or undisclosed alteration, and that in no single case could the medical records be fully reconciled with the descriptions, diagnoses, or histories published in the journal. (Godlee et al. 2011)

These disclosures, accompanied by a serious conflict of interest charge against Wakefield, led 10 of the original 13 study authors to retract support for the findings (Triggle 2010). In 2007, the UK General Medical Council (GMC) began an examination of professional misconduct. In 2010, the GMC ruled against Andrew Wakefield in the most severe way possible, by striking him from the medical register, thus ending his career as a physician in the UK (Meikle and Boseley 2010). That same year, the *Lancet* retracted the 1998 publication as unscientific and fraudulent (Wakefield et al. 2010). It would be hard to overestimate the damage this misinformation has caused in terms of public confidence in vaccines, not only in the UK but also in the United States and many other developed countries (Fig. 12.1).

However, the problem doesn't end there. A significant public outcry has also arisen against thimerosal, an ethyl mercury-containing material used in multi-dose vials of certain vaccines since the 1930s. To be clear, the MMR vaccine has never contained thimerosal, and the compound has been removed from all pediatric vaccines since 2001 (Kollaritsch and Rendi-Wagner 2012; Thimerosal in Vaccines 2015). Nonetheless, since the early 2000s, many groups have claimed, once again, an association with autism. Other than local, usually mild, hypersensitivity reactions, the Food and Drug Administration and the Institute of Medicine have not shown any harmful effects of thimerosal in vaccines (Kollaritsch and Rendi-Wagner 2012).

Paul Offit, writing in the *New England Journal of Medicine*, has said this:

> The notion that thimerosal caused autism has given rise to a cottage industry of charlatans offering false hope, partly in the form of mercury-chelating agents ... Although the notion that thimerosal causes autism has now been disproved by several excellent epidemiologic studies, about 10,000 autistic children in the United States receive mercury-chelating agents every year. Furthermore, this notion has diverted attention and resources away from efforts to determine the real cause or causes of the disorder. (Offit 2007)

Carson and Flood are even more critical in making this point:

> At tremendous expense to the public, there have now been numerous well-designed studies encompassing literally millions of children showing no association between vaccines and autism. Despite this, a recalcitrant anti-vaccine lobby either ignores or denies the data, or

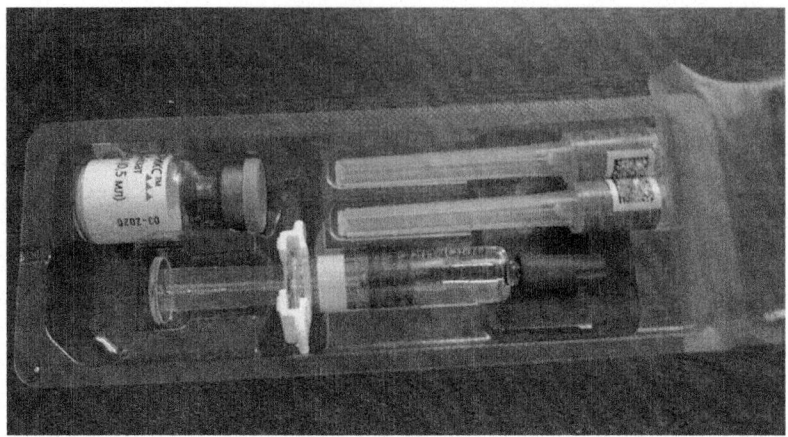

Fig. 12.1 Mumps-measles-rubella (MMR) vaccine, live, attenuated. (Commercial vaccine for the prevention of measles, epidemic mumps and rubella. The bottle with the green cap at upper left contains a freeze-dried preparation of attenuated viruses. At the upper right are two hypodermic needles with protective caps. At bottom is a pre-filled solvent syringe. Dctrzl: CC-BY-SA-4.0 Self-published work)

> keeps moving the target. Once the MMR (measles, mumps, rubella) vaccine was shown to have no association, concern was directed against thimerosal. Once claims against thimerosal were thoroughly debunked, the focus was on the number of vaccines. Now that several studies show no association there, we see the blame-shifting to aluminum adjuvants. This game of "whack-a-mole" to keep dispelling "any possible link" comes at a significant cost, with dollars and attention likely not being given to more fruitful avenues of scientific research. (Carson and Flood 2017a)

We could say much more about how vaccine misinformation has eroded public confidence, but the autism concern is significant. Education and advocacy play a vital role in alleviating these fears. However, there are specific, *truthful* claims about some vaccines that also contribute to public distrust and ethical concerns. The issue of *moral complicity* is at the heart of these objections.

As we discussed in the previous chapter on healthcare rights of conscience, moral complicity refers to the possible taint of moral guilt attached to a person by association with a moral wrong. In the context of vaccines, this issue comes up with the use of fetal tissue to develop and obtain certain vaccines for widespread use.[1] Such was the case with the rubella vaccine, which used a live (though attenuated) virus developed in tissue culture taken initially from aborted fetuses (Plotkin 1965). Today, two common cell lines currently exist that came from aborted fetal tissue: the WI-38 line, originally from lung tissue of a fetus aborted at 3 months gestation, (WI-38 2020) and the MRC-5 line, originally from lung tissue of a fetus aborted at

---

[1] Portions of this section first appeared in: Sullivan DM, Costerisan A. Complicity and Stem Cell Research: Countering the Utilitarian Argument. *Ethics & Medicine*. 2008;24(3):151–158. Used with permission.

14 weeks gestation (MRC-5 2020). The WI-38 line is still used for MMR vaccine production, while the MRC-5 line is used for Havrix® (Hepatitis A), Twinrix® (Hepatitis A/Hepatitis B), Varivax® (varicella), and Zostavax® (shingles) (Vaccine Excipient Summary 2020).

Some would make a strong case against using vaccines derived from aborted tissue, because of the idea of complicity with the original act, which those from a religious perspective would consider immoral. Vaccine *users* (physicians and their patients) are removed from the original act, however. The *intent* of those end-users has nothing to do with abortion, as they were not part of that initial decision. Furthermore, there seems to be a strong intuition that the passage of time reduces complicity to a morally repugnant act.

Nonetheless, some faith-based voices remain troubled by this apparent "moral taint." Catholic writer James Burtchaell derisively refers to the passage of time as a sort of "moral autoclave," as though it could sterilize the original evil act (Burtchaell 1989). Alexander Pruss seems to dismiss this approach as well:

> Thinking about the time elapsed since the evil was committed is one way of distancing oneself psychologically, but this may also involve self-deceit – a way of closing one's eyes to the evils of the past, similar to the way we close our eyes to starvation on other continents. (Pruss 2006)

Many current religious voices, however, while not minimizing moral concerns about the perceived evils of abortion, nonetheless maintain that widespread immunization is a desirable good for all of society collectively. Carson and Flood make this case well:

> We must not surrender the tremendous gains that vaccines have afforded us in the fight against infectious diseases. Christians and non-Christians alike who exercise their individual freedoms by refusing to vaccinate themselves or their children do so at the expense of their own and their neighbor's good. Especially when clustering together, these individuals increase the risk of both the return and spreading of several diseases, which threaten the health and lives of themselves and those unable to receive vaccines. For its part, Catholic social teaching entails a duty to vaccinate in order to protect the vulnerable. Individuals pursuing solidarity should act justly toward their neighbors and lovingly seek their good. (Carson and Flood 2017b)

How should pharmacists respond to these challenges? We suggest the following prudential approach: pharmacists should **affirm**, **inform**, and **empower**.[2]

Pharmacists should **affirm** the validity of patient concerns over any aggressively marketed pharmaceutical product. It is not wrong to harbor suspicions about an industry that has not always acted in the public's best interest. After all, suspicion itself is not evidence of ignorance, but a healthy manifestation of self-protection, or of parents' duty to protect their child. Such displays of caution are praiseworthy, and we should commend our patients for this. Furthermore, the Health Belief Model explains how personal beliefs and perceptions regarding vaccines strongly influence decisions regarding vaccination (Cheney and John 2013). If patient or caregivers'

---

[2] These insights specifically offered by PharmD candidate Joshua Pearson (class of 2021), who has also co-authored this chapter.

perceptions are roundly disregarded by pharmacists, any subsequent dialog regarding the risks and benefits of vaccines is likely to be unsuccessful at best and perceived as unsympathetic at worst.

The next step is to **inform.** Pharmacists should readily share information about vaccines and their side effects or possible links to various health concerns, including autism. However, a word of caution is important here: debunking myths on a personal level almost always backfires, as today's society balks at a paternalistic approach to education regarding vaccines. A more effective approach might be to sympathize with the patient by saying something like, "I have read about those concerns and take them seriously as well. This had led me to spend time looking into these claims. May I share the information that I have found?" This respects patient autonomy (by asking permission), shows genuine care, and demonstrates empathy that is the foundation for the pharmacist-patient relationship.

Some patients are concerned about adjuvants, such as aluminum, which can be found in trace amounts in vaccines. The reality, however, is that we likely absorb more aluminum from canned soft drinks in a year than can be found in an entire childhood vaccination schedule (Šeruga et al. 1994). Patients and parents should also consider that trace aluminum is present in deodorant, infant formula, and common over-the-counter remedies (including diaper ointments). These products have not had any measurable adverse effects on their users (Corkins 2019). As far as any other possible contaminants are concerned, vaccine manufacturers are routinely inspected, and each lot is tested for impurities; when these tests reveal a contaminant, the vaccine is immediately recalled (Vaccine Recalls 2020; Huang et al. 2010).

As far as mercury is concerned, it is worth pointing out to patients and parents that almost all vaccines have phased out any mercury-containing ingredients. Only certain multi-dose preparations of influenza vaccines still contain thimerosal as a preservative (Thimerosal and Vaccines 2018). Nonetheless, the ethyl mercury in thimerosal is organic, far different from the toxic form of mercury (methyl mercury) sometimes found in fish and industrial contaminants (Dórea et al. 2013). Furthermore, thimerosal is conjugated and easily removed from the body, while its toxic "cousin" is more apt to remain in the body and cause harm.

After affirming and informing, the final step is to **empower.** Using patients' protective instincts for themselves and their children, pharmacists can empower them to avoid infectious diseases. Parents should understand the concept of *herd immunity*, also known as "community immunity," the idea that if a majority of the population is immunized, this protects everyone (Fine et al. 2011). For example, herd immunity for measles occurs at a vaccine rate of 93–95% (Funk 2017). Parents accept vaccines for their children better when clinicians explain this (Betsch et al. 2017).

Finally, let us consider the honest ethical objections, usually from a religious perspective, to a perceived unethical source for some vaccines. For vaccines developed from aborted fetal remains, it is worth noting that the original purpose of the abortions was unrelated to that goal. Whatever reasons mothers chose to have an abortion, the decision to use the resulting tissue came after that fact. Seen in this light, this may confer a redemptive element onto an otherwise tragic event: the perceived death of a human person. The following are some insights from selected religious perspectives:

Roman Catholicism: the Pontifical Academy for Life encourages continued research to develop vaccines without the use of fetal tissue. Nonetheless, they promote the idea of using current vaccines for the sake of preserving life and loving one's neighbor (Luño 2006).

Protestant Christianity: One should take care that the pursuit of personal piety is not at the cost of neighborly love. The Bible exposes the hypocrisy of maintaining holy external appearances while ignoring the needs of others. According to the Christian scriptures, Jesus performed many healings on the Sabbath, in defiance of Pharisaical law (e.g., Matt. 12:9–14; Mark 1:21–28; Luke 13:10–17). Also, statements from several major Protestant denominations permit the use of vaccines from aborted fetal tissue due to the inability to change the past, the fact that no further abortions are needed to create the vaccines, and the positive prospect of individual and herd immunity. Using such vaccines is seen as making the best of our present circumstances despite past unethical practices (Ruijs et al. 2013; Best et al. 2016; Grabenstein 2013).

Buddhism: With specific regard to vaccines derived from fetal tissue, Buddhism emphasizes refraining from taking life and striving to preserve life. One must, therefore, carefully weigh the competing evils of vaccine development against the public danger associated with not vaccinating (Pelčić et al. 2016). In general, Buddhism is favorable to medicines not derived from animals since medical treatment is an act of mercy. Buddhist nuns employed variolation (an early form of vaccination) over one thousand years ago (Grabenstein 2013).

Islam: For Muslims concerned about the use of vaccines, the Prophet Mohammad (peace be upon him) reportedly said that Allah provided a cure for every disease.[3] Muslim nations knew about and practiced vaccination long before Western countries popularized the practice (The Dakar Declaration on Vaccines 2014). Devout Muslims may raise an additional concern regarding animal products, specifically of porcine origin, found in vaccines or utilized in the manufacturing process. However, the law of necessity allows Muslims to use these products when no alternative exists and when doing so fulfills the principles of preventing harm (*izalat aldharar*) and the promotion of public interest (*maslahat al ummah*) (Pelčić et al. 2016).

Judaism: While not explicitly addressed in the Torah or traditions, it is generally accepted by Jewish scholars that the utilization of vaccines follows the principle of *Pikuakh nefesh* – the act that saves lives (Turner 2017). *Kashrut* is the aggregate of Jewish dietary laws adhered to by some branches of Judaism. Foods derived from unclean animals such as pigs are considered to be non-kosher and thus are prohibited.[4] However, many authorities within the faith agree that this porcine prohibition

---

[3] Book of Medicine (Book 29, Hadith 20). *Sunan Abi Dawud 3874.* https://sunnah.com/abudawud/29/20

[4] Jewish Dietary Laws (Kashrut): Overview of Laws & Regulations. *Jewish Virtual Library.* 2020. www.jewishvirtuallibrary.org/overview-of-jewish-dietary-laws-and-regulations

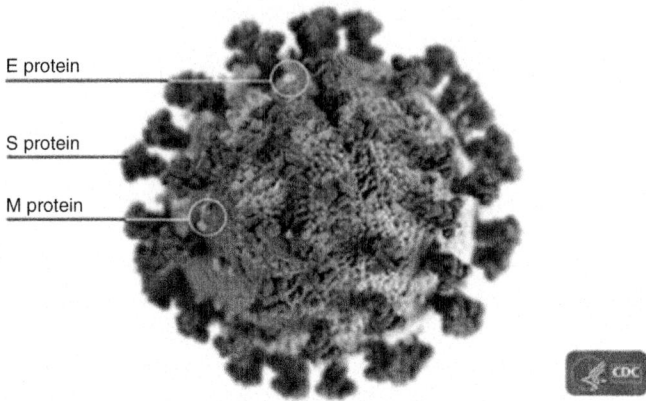

Fig. 12.2 COVID-19 Virus Illustration. (Public Domain: two employees of the Center for Disease Control created this graphic to show what the COVID-19 virus might look like)

applies strictly to orally-ingested substances. Non-porcine formulations of injectable vaccines are therefore preferable but not obligatory (Turner 2017).

A full analysis of religious views about vaccines is beyond the scope of this text. However, pharmacists have an interest in understanding the unique perspectives of their patients, which are often informed by religious beliefs. The subsequent advice that pharmacists offer to individual patients and the parents of small children should be reassuring and explanatory, but never overly zealous or pushy. Ultimately, the decision lies with the individual. Any perceived sense of compulsion will likely backfire.

Ethics During a Pandemic

We are writing this chapter midway through 2020, a unique and challenging time for healthcare and the world population in general. An infectious disease pandemic has sickened many, caused by severe acute respiratory syndrome coronavirus 2 (SARS-CoV-2) or, more simply, coronavirus disease 2019 (COVID-19). This outbreak first began in Wuhan, China in December 2019, but subsequently spread to more than 180 countries (Coronavirus (COVID-19) 2021) (Fig. 12.2).

When there is little pre-existing immunity to an infectious disease, it can spread worldwide and cause a pandemic (a global outbreak).[5] While challenging, the COVID-19 crisis is not the first pandemic, even during the last century. In 1918–1920, an outbreak of influenza A, subtype H1N1, popularly called the

---

[5] Coronavirus Disease 2019 (COVID-19): Situation Summary. Centers for Disease Control and Prevention Web site. www.cdc.gov/coronavirus/2019-ncov/cases-updates/summary.html#. Published 2021. Accessed 2/15/2021.

"Spanish Flu," took the lives of 50–100 million people across the globe.[6] More recently, in the spring of 2009, another new H1N1 strain appeared, first detected in the U.S. Often called the "Swine Flu," it rapidly spread around the world, and by mid-2010, it had led to over 12,000 deaths in the U.S. alone.

It is beyond the purpose of this text to go deeply into all of the problems such crises create. We will consider just one: the ethical distribution of medical care when hospital facilities are overwhelmed. During the early months of the COVID-19 crisis, many hospitals and healthcare systems developed policies and protocols to allocate scarce resources. Since patients with COVID-19 infection may develop respiratory failure, there was great concern that this emergency would create a demand for mechanical ventilators, medications used to support mechanical ventilation, and other critical resources that would outstrip the supply. Such scarcity would necessitate tough policy decisions over who to ventilate and who to leave unventilated. Clinical facilities should only implement crisis allocation policies if: (1) critical care capacity has been, or shortly will be, exceeded despite taking all appropriate steps to increase capacity, and (2) a regional-level authority has declared an emergency (White and Lo 2020).

As we have repeatedly stressed throughout this text, the traditional rules of medical ethics date back to more than two millennia. These include *beneficence,* the idea that healthcare professionals will always act in the best interests of their patients and *non-maleficence*, the avoidance of harm. Another key principle is *distributive justice*, the fair and equitable distribution of the benefits of healthcare, regardless of age, gender, social class, ethnicity, sexual orientation, ability to pay, religion, handicap, or any other medically non-relevant trait (Beauchamp and Childress 2019; Edelstein 1943). In the modern era, we have added *autonomy*, which allows patients and their surrogates to make their own decisions based on personal values, which is the basis for informed consent (Jonsen et al. 2015; Beauchamp 2011).

In a resource-limited environment, however, such as during a significant healthcare crisis, it will not always be possible to provide the most definitive and beneficial treatment to all patients. In such cases, healthcare ethics must utilize a different set of rules, based on the idea of *triage*. The word triage comes from the French *trier*, which means "to sort." Initially used in a military combat setting, the triage process assigns a category to individuals within a large group of causalities, to enable rapid decision-making to save the most lives (Mitchell 2008). This approach means that a form of rationing may be necessary:

> The fundamental point of triage is the following: not everyone who needs a particular form of health care, such as medicine, therapy, surgery, transplantation, or intensive care bed, can gain immediate access to it. Triage systems are designed to assist allocation decisions in this regard. These decisions are more difficult when a condition is life-threatening and the scarce resource potentially life-saving. In life-threatening conditions, the question can become: "Who shall live when not everyone can live?" (Aacharya et al. 2011)

---

[6] 1918 Pandemic (H1N1 virus). Centers for Disease Control and Prevention Web site. www.cdc.gov/flu/pandemic-resources/1918-pandemic-h1n1.html. Published 2019. Accessed 11/29/2020.

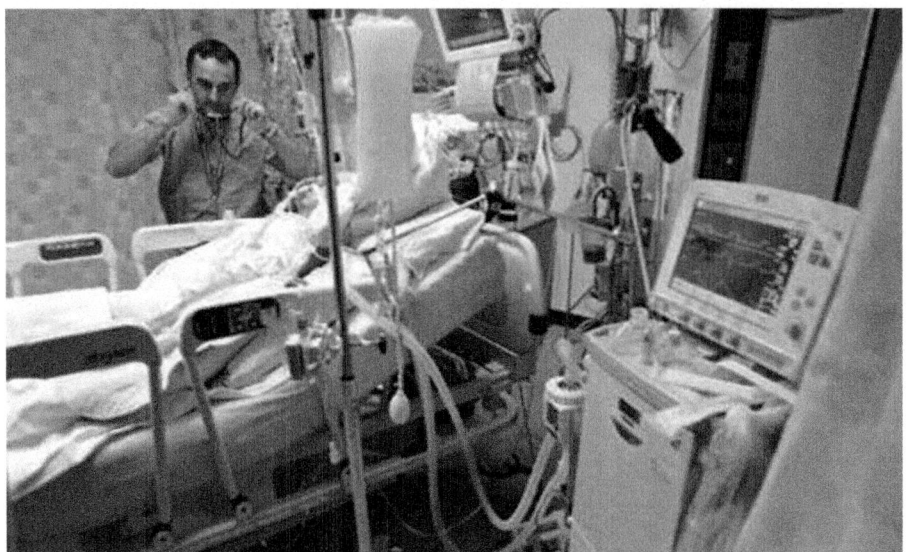

Fig. 12.3 A mechanical ventilator in use for an ICU patient. (By Rcp.basheer - Own work, CC BY-SA 3.0, https://commons.wikimedia.org/w/index.php?curid=16721005)

The primary purpose of a triage protocol is to save the most lives, which, of course, comes from utilitarian ethics, seeking to maximize the best outcomes for the highest number of people (Mill 1859). Ventilator allocation guidelines from the New York State Task Force rely on this understanding (Fig. 12.3):

> The Guidelines define survival by examining a patient's short-term likelihood of surviving the acute medical episode and not by focusing on whether the patient may survive a given illness or disease in the long-term (e.g., years after the pandemic). Patients with the highest probability of mortality *without* medical intervention, along with patients with the smallest probability of mortality *with* medical intervention, have the *lowest* level of access to ventilator therapy. Thus, patients who are most likely to survive *without* the ventilator, together with patients who will most likely survive *with* ventilator therapy, increase the overall number of survivors. (Ventilator Allocation Guidelines 2015)

However, devising an equitable and just protocol means more than merely maximizing the number of patients who survive to hospital discharge. Other criteria may be employed, as laid out in a protocol from Maryland:

- Prospects for Short-Term Survival: the most straightforward measure of whether a patient will benefit from intensive care is whether that patient survives as a consequence of this care.
- Prospects for Long-Term Survival: also related to how much benefit intensive care produces is the patient's prospects for survival after discharge. Although essential, placing too high a priority on the criterion of long-term survival may, in certain circumstances, further disadvantages people who already face systematic hardships (i.e., this may be discriminatory).

- Pregnancy: Gives a slight preference for a woman in respiratory failure with a healthy fetus.
- Fair Chance: If patients have equal priority based on allocation criteria, then the allocation is based on 1) first-come, first-served or 2) a lottery (Biddison et al. 2019).

Another principle often appears in surveys of community members, and that is the idea of preserving those who have more life-years left, in other words, younger individuals (Biddison et al. 2014). Some model protocols attempt to address this idea. However, these may run afoul of civil rights law, which forbids discrimination based on age.[7] To ethically allocate precious resources during a healthcare crisis therefore requires careful advance planning, which hopefully has already taken place with well thought-out protocols long before a crisis occurs.

The need for ethical thinking during a pandemic also applies to the drug supply chain. Initial reports during the COVID-19 pandemic pointed to the potential efficacy of the combination of hydroxychloroquine and azithromycin (HC + AZ) in treating acutely ill patients (Kim et al. 2020). Soon after the publication of these reports and the subsequent media attention, wholesalers, health systems, pharmacies, and practitioners began a mad rush to acquire these drugs. However, hoarding these agents could impact the availability of these medications for patients who need them for malarial prophylaxis or treatment, as well as for autoimmune diseases such as lupus and rheumatoid arthritis. Some states, therefore, passed emergency rules to restrict the use of chloroquine and hydroxychloroquine without a documented diagnosis code, including for the treatment of laboratory-confirmed COVID-19 cases.[8] At the time of writing this chapter, clinical studies have called into question the effectiveness of HC + AZ. Yet this problem dramatically highlights the intersection between distributive justice and pharmacy law.

Planning for a resource shortage is essential, but it is only preliminary. As of this writing, none of the myriad triage and allocation protocols have been invoked in the United States or Canada. Nonetheless, crises generate considerable anxiety, and this may sometimes distort our thinking. As a given crisis mounts, it may be tempting for some clinicians to shift into a more utilitarian mode of reasoning, and begin to "ration" resources. However, the usual principles of medical ethics still apply – beneficence, non-maleficence, distributive justice, and autonomy – up until resource scarcity triggers a policy shift towards triage. Therefore, it is imperative that the circumstances that invoke an allocation policy be well understood beforehand. No matter how dire the circumstances, there is no justification for changing time-honored ethical rules on the fly (for the latest developments, see the 'Pandemic Update' at the end of this chapter).

---

[7] OCR Issues Bulletin on Civil Rights Laws and HIPAA Flexibilities That Apply During the COVID-19 Emergency. *Health and Human Services: Office of Civil Rights.* 2020. www.hhs.gov/about/news/2020/03/28/ocr-issues-bulletin-on-civil-rights-laws-and-hipaa-flexibilities-that-apply-during-the-covid-19-emergency.html. Published 3/28/2020.

[8] Emergency Rule for Dispensing Chloroquine and Hydroxychloroquine. State of Ohio Board of Pharmacy Web site. https://med.ohio.gov/. Published 2020. Updated 3/26/2020. Accessed 2/15/2021.

The Ethics of Unproven Treatments

As we saw in Chap. 7 of this text, health and wellness promotion is an important role for pharmacists, and much of that involves patient education. Writing now as one of the authors of this text (DCA), I once practiced at a Veterans Administration community-based outpatient clinic. One of my patients brought in an over-the-counter herbal cold remedy to ask me if it might interfere with any of his current medications and to inquire as to its possible benefit. One of the ingredients was 'saw palmetto,' touted to "maintain prostate health." I pointed this out to him and asked, "If you asked me for a cold remedy, and I said, 'Here, take this prostate medicine,' what would you say?" The patient blinked a couple of times and answered, "I'd call you a *quack*."

In the above example, the patient recognized that an ingredient in the cold remedy had no possibility of benefit for that indication. He understandably concluded that it was not just unproven, but illegitimate. However, there were other ingredients in the remedy, including echinacea (albeit in a subtherapeutic dose). Echinacea has been claimed to shorten the duration and lessen the severity of colds. But randomized clinical trials have yet to offer substantial evidence of efficacy (Turner et al. 2005). This example leads us to two questions: (1) what about treatments that are potentially beneficial, but unproven, and (2) what is quackery?

As mentioned earlier, during the recent COVID-19 pandemic, preliminary reports and public statements led to sensationalism regarding a wide range of possible treatments, including HC + AZ (Kim et al. 2020; Gautret et al. 2020; Grein et al. 2020; Cao et al. 2020). Was the early release of data and the subsequent grasping of unproven treatments an example of medical quackery? Let's define these terms (Fig. 12.4).

As with the HC + AZ example, some treatments are potentially therapeutic but not yet proven. As we discussed in Chap. 6, a state of *clinical equipoise* exists when we genuinely do not yet know if one treatment is better than another, or if a particular treatment does more good than harm (Kukla 2007). Using an unproven treatment that may have the potential for therapeutic benefit does not constitute medical quackery. There are also many treatments where the benefit is difficult to prove, but they are not particularly harmful (Ulbricht et al. 2018; Ulbricht and Ko 2017). This does not represent medical quackery either.

True *medical quackery* occurs when one or more of the following conditions are met:

1. A treatment has no potential for benefit.
2. The known harm of a treatment is significantly greater than any potential benefit.
3. Those promoting the treatment are being deliberately deceptive, usually for profit.

Therefore, the HC + AZ combination for treating COVID-19 represented clinical equipoise. It later turned out that the treatment was not as effective as initially hoped; in fact, outcomes were slightly worse with the treatment (Magagnoli et al. 2020). On the other hand, nebulized hydrogen peroxide as a proposed treatment for

Fig. 12.4 Illustration from *Nostrums and Quackery*, published by the American Medical Association (1914). (Source book page: https://archive.org/stream/nostrumsquackery01ameruoft/nostrumsquackery01ameruoft#page/n438/mode/1up, No restrictions, https://commons.wikimedia.org/w/index.php?curid=43319723)

COVID-19 has been recently exposed as a form of medical quackery (Caulfield 2020; Jordan et al. 1991; Garcia 2020).[9] William Jarvis perhaps said it best: "Quackery not only harms people, it undermines the scientific enterprise and should be actively opposed by every scientist." (Jarvis 1992).

Upton Sinclair's *The Jungle*, the classic muckraking exposé of the meatpacking industry, spurred the passage of the Pure Food and Drug Act of 1906 that eventually created the U.S. Food and Drug Administration (FDA). During the 1930s, the sale of sulfanilamide elixir, with the poison diethylene glycol as its solvent, led to over 100 deaths. This public scandal led to the passage of the Food, Drug, and Cosmetic Act of 1938, which created the classification of prescription-only medicines (Akst 2013). Subsequent laws and amendments required medications to prove not only their safety but also their effectiveness before marketing (Kurian and Harahan 1998). Nonetheless, even with a heavily regulated pharmaceutical industry, many dubious and worthless treatments are still available for patients to purchase. Having a given medication available in a pharmacy provides the remedy with a patina of authority. Pharmacists should, therefore, carefully consider the products they sell.

The ethics of all this may get complicated. In the case of bogus treatments for COVID-19, many patients who put their trust in them believed that such products would make them "immune" to the virus. Such thinking delayed diagnosis and treatment and led to further spread of the virus while patients were unaware of being

---

[9] Questionable methods of cancer management: Hydrogen peroxide and other "hyperoxygenation" therapies. *CA: a cancer journal for clinicians*. 1993;43(1):47–55.

contagious. Delaying treatment may potentially cause harm, violating non-maleficence. Similarly, promoting a treatment with no possibility of benefit violates beneficence, since the purveyors are not using their medical knowledge for a patient's best interest.

Furthermore, it violates distributive justice to take advantage of desperate patients with incurable diseases who are not able to evaluate false claims properly. Fraudulent treatments may seem less expensive than traditional healthcare, but delaying treatment until the later stages of a disease may be a costly decision. For complementary, alternative, or integrative medicine (CAM) therapies, the health benefit is difficult to prove. Nonetheless, these treatments give patients choices and therefore promote autonomy (Ulbricht and Ko 2017). Conscientious pharmacists must endeavor to fully educate their patients so that they utilize CAM therapies with informed consent. It is also necessary to educate and inform patients of the risks and benefits of unproven treatments such as HC + AZ for COVID-19, or, more commonly, supplements with false or exaggerated claims of benefit.

Part of pharmacy professionalism is having the ability to make sense of medical information and communicate this to our patients in an understandable and empathetic way. This commitment means warning them of the dangers of fraudulent medications and quackery. It also requires cautious counseling regarding questionable treatments and not overly promoting unproven ones. The following table summarizes these different categories (Table 12.2).

Pandemic Update

We write this section in early March, 2021, as an update to our earlier discussion on the ethical dilemmas presented by the COVID-19 pandemic. 28 million people in the United States have been diagnosed with the disease since the crisis began, nearly one year ago. Case numbers skyrocketed in the weeks following the 2020 Christmas and New Year holidays, to a high of 214,000 daily cases on January 8, 2021. These numbers plummeted to 28,000 by February 21, 2021, finally lower than the summer peak of 2020. On February 20, 2021, the U.S. passed a grim milestone, with more than 500,000 confirmed deaths due to COVID-19,[10, 11, 12] more American fatalities than the combined toll of both world wars and the war in Vietnam. The following chart graphically shows these developments (Fig. 12.5):

---

[10] US coronavirus map: Tracking the outbreak. *USA Today.* https://covid.cdc.gov/covid-data-tracker/#trends_dailytrendscases. Published 2/21/2021. Accessed 2/24/2021.

[11] COVID-19. *Centers for Disease Control and Prevention.* (2/24/2021). www.cdc.gov/coronavirus/2019-ncov/index.html. Published 2/21/2021.

[12] Huang P. 'A Loss To The Whole Society': U.S. COVID-19 Death Toll Reaches 500,000. *National Public Radio.* www.npr.org/sections/health-shots/2021/02/22/969494791/a-loss-to-the-whole-society-u-s-covid-19-death-toll-reaches-500-000. Published 2/22/2021.

Table 12.2 Categories and Examples of Treatments with Varying Benefits or Harms[a]

Category	Therapy
Fraudulent treatments or medical quackery Known to have no benefit or harmful	Laetrile (Milazzo et al. 2007) Colloidal silver (for COVID-19) Chiropractic spinal manipulation as a replacement for vaccination (Gleberzon et al. 2013; Barrett 2015) Intravenous hydrogen peroxide (Jordan et al. 1991; Hirschtick et al. 1994)
Questionable treatments Limited or no evidence of benefit or harm	Homeopathy (Ernst 2010) Naturopathy (Cooley et al. 2009) Acupuncture (Wang et al. 2008) Ayurveda (Savaliya et al. 2010; Saper et al. 2008) Magnetic bracelets for arthritis pain (Pittler et al. 2007)
Unproven treatments Benefit or harm unknown, benefit theoretically possible	Hydroxychloroquine sulfate with or without azithromycin for COVID-19 (Gautret et al. 2020; Geleris et al. 2020) Vitamin C for the common cold (Hemilä and Chalker 2013) Extended thromboprophylaxis after medical illness (Spyropoulos et al. 2018) Chiropractic spinal manipulation for low back pain (Walker et al. 2010)

[a]Not intended as an exhaustive list; these are examples to spur further study (see references)
[b]FDA and FTC Warn Seven Companies Selling Fraudulent Products that Claim to Treat or Prevent COVID-19. *FDA News Release.* 2020. www.fda.gov/news-events/press-announcements/coronavirus-update-fda-and-ftc-warn-seven-companies-selling-fraudulent-products-claim-treat-or. Published 3/9/2020

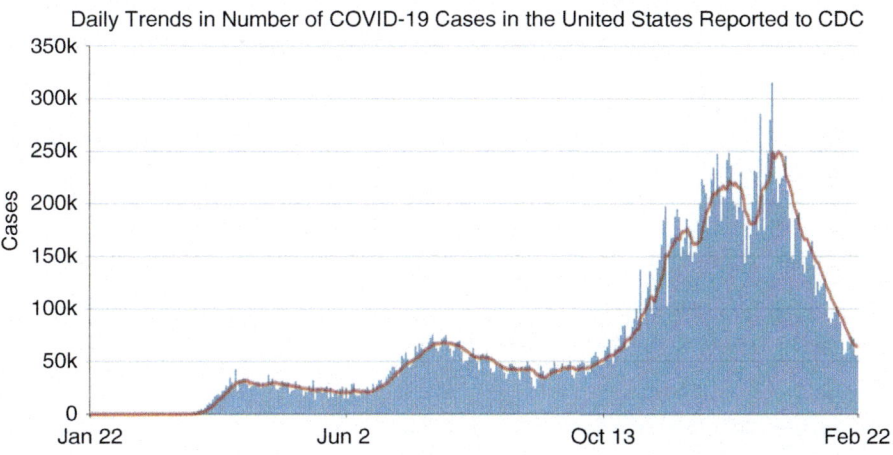

Fig. 12.5 Daily Trends (CDC) (COVID-19. *Centers for Disease Control and Prevention.* (2/24/2021). www.cdc.gov/coronavirus/2019-ncov/index.html. Published 2/21/2021)

The worrisome numbers at the end of 2020 caused huge concerns among public health officials and threatened to overwhelm healthcare facilities, especially in heavily populated states such as New York and California. What caused improvement in 2021? Former CDC director Tom Frieden said it is because of better adherence to social distancing guidelines, "It's what we're doing right: staying apart, wearing masks, not traveling, not mixing with others indoors."[13]

However, in order to see a long-term reduction in cases, implementing COVID-19 vaccines appears to be one of the best strategies.[14] For this reason, "Operation Warp Speed" (OWS) was initiated in May of 2020, an ambitious federal program to accelerate vaccine development, based on a strategic partnership between the Department of Health and Human Services (DHHS) and the Department of Defense, along with private companies (Slaoui and Hepburn 2020).

Of the many types of COVID-19 vaccines under development, this brief discussion will mention three. The first vaccine to receive Emergency Use Authorization (EUA) by the FDA was by the Pfizer Corporation and its German partner BioNTech, with the EUA granted on December 11, 2020.[15] The next was the Moderna vaccine, which received its EUA on December 18, 2020.[16] Johnson & Johnson's vaccine, developed in collaboration with Janssen Biotech, received the same approval on February 27, 2021.[17]

The Pfizer/BioNTech and Moderna vaccines use the newer technology of messenger RNA (mRNA) to create viral immunity. This is accomplished by injecting a small portion of the coronavirus mRNA that codes for an outer shell or "spike" protein. The mRNA then causes the body's own cells to recreate that protein and stimulate an immune response to it.[18] The Johnson & Johnson vaccine uses a harmless, non-COVID viral vector. This stimulates immunity in the recipient, once again against the COVID-19 spike protein.[19]

In Phase 3 clinical trials of these vaccines, the results have been surprisingly robust. The Pfizer/BioNTech product requires two injections, three weeks apart, but is 95% effective in preventing symptomatic infection after the second dose. The

---

[13] Tinker B. Precautions not vaccines are helping the current decline in case rates, former CDC director suggests. *CNN World News.* Published 2/14/2021.

[14] Some material on the ethics of COVID-19 vaccines was first written by Rachel Wofthoff (PharD candidate, Class of 2023), who joins us as co-author of this section. Her original comments are available at: www.cedarville.edu/disruptive-healthcare/beyond-safety-and-efficacy-ethics-and-the-covid-vaccine. Used with permission.

[15] Pfizer-BioNTech COVID-19 Vaccine. *US Food and Drug Administration.* www.fda.gov/emergency-preparedness-and-response/coronavirus-disease-2019-covid-19/pfizer-biontech-covid-19-vaccine. Published 12/11/2020.

[16] Moderna COVID-19 Vaccine. *US Food and Drug Administration.* www.fda.gov/emergency-preparedness-and-response/coronavirus-disease-2019-covid-19/moderna-covid-19-vaccine. Published 12/18/2020.

[17] Coronavirus (COVID-19) Update: Janssen COVID-19 Vaccine. *US Food and Drug Administration.* https://www.fda.gov/emergency-preparedness-and-response/coronavirus-disease-2019-covid-19/janssen-covid-19-vaccine. Published 2/27/2021.

[18] Understanding mRNA COVID-19 Vaccines. *Centers for Disease Control and Prevention.* www.cdc.gov/coronavirus/2019-ncov/vaccines/different-vaccines/mrna.html. Published 12/18/2020.

[19] Understanding Viral Vector COVID-19 Vaccines. *Centers for Disease Control and Prevention.* www.cdc.gov/coronavirus/2019-ncov/vaccines/different-vaccines/viralvector.html. Published 1/5/2021.

Moderna vaccine, which requires two shots at a four-week interval, has an efficacy of 94%. The Johnson & Johnson one-dose vaccine is only 72% effective, but confers 85% protection against severe disease. To date, no deaths have been reported from COVID-19 infection in individuals receiving a COVID-19 vaccine during clinical trials.[20]

As discussed earlier in this chapter, so-called "herd immunity" is thought to be the best strategy to mitigate the toll of the crisis and return to normal. Surprisingly, even those previously infected with COVID-19 may not have the robust protection offered by current vaccines.[21, 22] Best estimates for achieving herd immunity will require vaccinating at least 70% of the population (Kadkhoda 2021). Based on an estimated U.S. population of 330 million, that would call require more than 230 million people to participate, an enormous task. Yet therein lies the biggest problem, for many members of the public have expressed doubt and uncertainty about COVID-19 vaccines, and some have ethical concerns. One major question seems to be this: was their deployment simply too rapid to know about unknown and unwanted potential side effects?

Under OWS, COVID-19 vaccines were developed in record time, yet the FDA has closely monitored them and assures their safety. Though the timeline has been greatly accelerated, no safety steps were omitted. An official DHHS summary explained it this way:

> [V]accine development began in January, phase 1 clinical studies in March, and the first phase 3 trials in July ... [W]e must accelerate vaccine program development without compromising safety, efficacy, or product quality. Clinical development, process development, and manufacturing scale-up can be substantially accelerated by running all streams, fully resourced, in parallel. Doing so requires taking on substantial financial risk, as compared with the conventional sequential development approach. OWS will maximize the size of phase 3 trials (30,000 to 50,000 participants each) and optimize trial-site location by consulting daily epidemiologic and disease-forecasting models to ensure the fastest path to an efficacy readout. Such large trials also increase the safety data set for each candidate vaccine. (Slaoui and Hepburn 2020)

Further adding to this assurance with the EUA now in place for the Pfizer and Moderna vaccines, there will soon be an even larger data set from the immunized U.S. population, now estimated at over 66 million people who have received at least one injection.[23]

Another reason for vaccine hesitancy is the fear that somehow mRNA vaccines are a form of "gene editing" that can modify a person's genetic code. Though mRNA

---

[20] Branswell H. Comparing the Covid-19 vaccines developed by Pfizer, Moderna, and Johnson & Johnson. *Stat News*. www.statnews.com/2021/02/02/comparing-the-covid-19-vaccines-developed-by-pfizer-moderna-and-johnson-johnson/. Published 2/2/2021.

[21] Quast I, Tarlinton D. B cell memory: Understanding COVID-19. *Immunity*. Published 1/23/2021.

[22] Skelly DT, Harding AC, Gilbert-Jaramillo J, et al. Vaccine-induced immunity provides more robust heterotypic immunity than natural infection to emerging SARS-CoV-2 variants of concern. *Research Square*. Published 2/9/2021.

[23] COVID-19 Vaccinations in the United States. *Centers for Disease Control and Prevention*. 2021. https://covid.cdc.gov/covid-data-tracker/#vaccinations. Published 2/24/2021.

technology is relatively new, it has been used before, and has been studied in vaccine trials for SARS-CoV-1, the Zika virus, rabies, and cytomegalovirus. Furthermore, there is no possibility that the COVID-19 vaccines can modify DNA, because they do not enter into the nuclei of cells, and are quickly metabolized and eliminated after inducing immunity. The same is true of viral-vector vaccines such as the one by Johnson & Johnson.

As we discussed in the earlier section on vaccines, there remains a concern about the use of fetal tissue cell cultures (derived from aborted fetuses) in vaccine development, of particular concern to religious conservatives. However, of the first two COVID-19 vaccine candidates from Moderna and Pfizer, neither one uses fetal tissue cell cultures in their actual production (though fetal cell cultures were used in early testing) (Corbett et al. 2020).[24] Some other vaccine candidates, such as the Johnson & Johnson product, have used fetal cell lines in both early testing and in subsequent production (Wong 2006; Tostanoski et al. 2020). Current guidelines from many religious authorities recommend the use of such vaccines in spite of these concerns, based on the utilitarian idea that widespread immunization resulting in herd immunity is a desirable goal (see earlier discussion on moral complicity).[25]

In terms of the science, OWS has been a historical success in its development of effective COVID-19 vaccines in such a short time. However, supply and distribution issues have caused disappointing delays in getting the vaccines to all Americans who need or want them. This is therefore a perfect case study on how distributive justice comes into play, and the debate centers on the prioritization of who should get vaccines first.

Ethically, at least three goals are in view: reducing overall morbidity and mortality, minimizing disease effects on the economy, and eliminating, as much as possible, healthcare disparities that disadvantage certain minority groups. At present, these priorities target the elderly, those with pre-existing comorbidities, and essential healthcare workers and first-responders (EMS, fire departments, police, etc.), as well as providers of essential services, such as postal employees, grocery personnel, and others (Gupta and Morain 2020). Distribution will then move from highest to lowest risk, though significant debate remains about how these risks should be determined. Many states have elected to prioritize vaccination based on age, vocation (e.g., school teachers), or the presence of chronic medical conditions. However, many have raised concerns about the limited vaccine availability among minority communities, who disproportionately have experienced negative impacts from the pandemic, including higher mortality rates.[26] Children will most likely be the last to

---

[24] Vogel AB, Kanevsky I, Che Y, et al. A prefusion SARS-CoV-2 spike RNA vaccine is highly immunogenic and prevents lung infection in non-human primates. *bioRxiv*. Published 9/8/2020.

[25] Physician Support for Ethical Vaccines. Christian Medical and Dental Association Web site. https://cmda.org/article/physician-support-for-ethical-vaccines/. Published 2020. Accessed 2/16/2021.

[26] Ndugga N, Pham O, Hill L, Artiga S, Mengistu S. Latest Data on COVID-19 Vaccinations Race/Ethnicity. *Kaiser Family Foundation*. www.kff.org/coronavirus-covid-19/issue-brief/latest-data-on-covid-19-vaccinations-race-ethnicity/. Published 2/18/2021.

receive the vaccines, since they tend to have the least severe effects from COVID-19 and were not included in clinical trials, though such studies are now in progress (Couzin-Frankel 2021). Regardless of the allocation strategies employed, more and more people should have access to vaccines as the supply increases.

During the recent surge of COVID-19 cases, intensive care units in hospitals have been overwhelmed. Many hospitals across the nation were able to mitigate the crisis through unprecedented measures, such as opening up unused hospital wings to increase critical care capacity (White and Lo 2021). As we discussed earlier, triage and resource allocation protocols are in place in most medical care facilities to govern the best utilization of limited resources. Surprisingly, because of creative mitigation strategies, such measures have almost never invoked. One exception was a brief recent period in several California cities (notably Los Angeles), where "crisis standards of care" were in place. This situation was ameliorated by the decline in case rates since the high in early January.[27, 28]

The COVID-19 pandemic is a perfect example of how healthcare systems can never be fully prepared for every ethical dilemma, yet the pharmacy profession can play a key role in finding solutions. In early February of 2021, the federal government announced that it would send one million vaccine doses to about 6500 retail pharmacies, with the eventual goal of getting vaccines directly to 40,000 drugstores and grocery stores across the nation. This represents a shift in strategy from placing vaccination sites in large clinical facilities and sports arenas, to instead utilizing smaller local settings. This could speed up vaccine distribution, especially in underserved communities (Robbins and Weiland 2021).

Conclusion

This chapter examined specific critical matters not previously discussed. We began with the ethical issues related to vaccines, considering how to deal with misinformation and genuine moral objections. Next, we turned to the ethics of pandemics and examined the difficult triage decisions that might be necessary when resources become scarce. We then discussed the use of fraudulent, questionable, and unproven treatments and how to ethically respond to each. This chapter concluded with a Pandemic Update, providing a further analysis of the ethics of dealing with the 2020–2021 COVID-19 pandemic, with an emphasis on the equitable distribution of vaccines against the disease. In the next chapter, we will put the reader to work analyzing clinical case studies and policy scenarios.

---

[27] Zaucha A. 'It's a Desperate Time': Crush of Covid-19 Patients Strains U.S. Hospitals. *The Wall Street Journal.* www.wsj.com/articles/its-a-desperate-time-crush-of-covid-19-patients-strains-u-s-hospitals-11609721551. Published 1/3/2021.

[28] Howard M. COVID-19 crisis in Los Angeles: Why activating 'crisis standards of care' is crucial for overwhelmed hospitals. *The Conversation.* https://theconversation.com/covid-19-crisis-in-los-angeles-why-activating-crisis-standards-of-care-is-crucial-for-overwhelmed-hospitals-152706. Published 1/6/2021.

Key Terms

- vaccination
- thimerosal
- moral complicity / moral taint
- herd immunity
- COVID-19
- H1N1
- triage / allocation
- medical quackery
- clinical equipoise
- CAM therapies
- Operation Warp Speed
- Emergency Use Authorization

Questions for Review

1. Why could we not repeat Edward Jenner's famous experiment with a smallpox vaccine today?
2. Summarize the underlying story behind the common concern about vaccines and autism.
3. Define moral complicity. How does this idea come into play with vaccines derived from aborted fetal tissue?
4. Summarize the three-part approach to vaccine counseling recommended in this chapter.
5. How does the idea of triage differ from the usual approach to clinical ethics?
6. Is it ethical to stock questionable and unproven medications or treatments in a community pharmacy? How should pharmacists advise their patients in using these?
7. During a pandemic, if you were directing vaccine distribution in your state and the supply was limited, who would receive the top priority to receive it? Give an ethical justification for your answer.

For Further Reading

Elena Conis: *Vaccine Nation: America's Changing Relationship with Immunization.* University of Chicago Press, 2002. Elena Conis is a historian and gives an even-handed account of vaccine development and its many controversies, beginning in the 1960s. She explores the politics, economics, and social factors that inform both vaccine advocates and vaccine skeptics.

Donal O'Mathúna and Walt Larimore: *Alternative Medicine: The Christian Handbook.* Zondervan, 2006. Donal O'Mathuna is a lecturer in healthcare ethics in Dublin, Ireland. Walt Larimore, a family physician, is also a popular author, educator, and medical journalist. Despite a distinctly faith-based format, this is a comprehensive resource for anyone on the complex array of alternative and complementary therapies. Extensive cross-references make this easy to use.

References

Aacharya RP, Gastmans C, Denier Y. Emergency department triage: An ethical analysis. BMC Emerg Med. 2011;11:16.

Akst J. The elixir tragedy, 1937. The Scientist Magazine. 2013. www.the-scientist.com/foundations/the-elixir-tragedy-1937-39231. Published 5/31/2013.

Barrett S. Chiropractors and Immunization. Chirobase. 2015. https://quackwatch.org/chiropractic/dd/chiroimmu/. Published 3/10/2016.

Beauchamp TL. Informed consent: Its history, meaning, and present challenges. Camb Q Healthc Ethics. 2011;20(4):515–23.

Beauchamp TL, Childress JF. Principles of biomedical ethics. 8th ed. New York: Oxford University Press; 2019.

Best MM, Sarvaananda S, Martin JM, White PG, Martin ML. Spiritual care services in emergency medicine. In: Diversity and Inclusion in Quality Patient Care. Springer; 2016:83–100.

Betsch C, Böhm R, Korn L, Holtmann C. On the benefits of explaining herd immunity in vaccine advocacy. Nat Hum Behav. 2017;1(3):1–6.

Biddison ELD, Gwon H, Schoch-Spana M, et al. The community speaks: Understanding ethical values in allocation of scarce lifesaving resources during disasters. Ann Am Thoracic Soc. 2014;11(5):777–83.

Biddison ELD, Faden R, Gwon HS, et al. Too many patients: A framework to guide statewide allocation of scarce mechanical ventilation during disasters. Chest. 2019;155(4):848–54.

Burtchaell JT. The giving and taking of life: Essays ethical. Notre Dame: University of Notre Dame Press; 1989.

Cao B, Wang Y, Wen D, et al. A trial of lopinavir–ritonavir in adults hospitalized with severe Covid-19. N Engl J Med. 2020;

Carson PJ, Flood AT. Response to open peer commentaries on Catholic social teaching and the duty to vaccinate. Am J Bioeth. 2017a;17(4):W1–3.

Carson PJ, Flood AT. Catholic social teaching and the duty to vaccinate. Am J Bioeth. 2017b;17(4):36–43.

Caulfield T. Pseudoscience and COVID-19-we've had enough already. Nature. 2020. www.nature.com/articles/d41586-020-01266-z. Published 4/17/2020.

Cheney MK, John R. Underutilization of influenza vaccine: A test of the health belief model. SAGE Open. 2013;3(2)

Chiswick ML. Jenner, romanticism, and research. Arch Dis Child. 1996;74(1):1.

Cooley K, Szczurko O, Perri D, et al. Naturopathic care for anxiety: A randomized controlled trial ISRCTN78958974. PLoS One. 2009;4(8)

Corbett KS, Edwards DK, Leist SR, et al. SARS-CoV-2 mRNA vaccine design enabled by prototype pathogen preparedness. Nature. 2020;586(7830):567–71.

Corkins MR. Aluminum effects in infants and children. Pediatrics. 2019;144(6)

Coronavirus (COVID-19). Centers for Disease Control and Prevention Web site. www.cdc.gov/coronavirus/2019-ncov/. Published 2021. Accessed 2/15/2021.

Couzin-Frankel J. As trials ramp up, doctors stress need to vaccinate kids against COVID-19. Science Magazine. 2021. www.sciencemag.org/news/2021/02/trials-ramp-doctors-stress-need-vaccinate-kids-against-covid-19. Published 2/23/2021.

Deer B. Revealed: MMR research scandal. The Sunday Times. 2004. www.thetimes.co.uk/article/revealed-mmr-research-scandal-7ncfntn8mjq. Published 2/22/2004.

Dórea JG, Farina M, Rocha JB. Toxicity of ethylmercury (and thimerosal): A comparison with methylmercury. J Appl Toxicol. 2013;33(8):700–11.

Edelstein L. The Hippocratic oath, text, translation and interpretation. Baltimore: The Johns Hopkins press; 1943.

Ernst E. Homeopathy: What does the "best" evidence tell us? Med J Aust. 2010;192(8):458–60.

Fine P, Eames K, Heymann DL. Herd immunity: A rough guide. Clin Infect Dis. 2011;52(7):911–6.

Funk S. Critical immunity thresholds for measles elimination. Centre for the Mathematical Modelling of Infectious Diseases. 2017. www.who.int/immunization/sage/meetings/2017/october/2._target_immunity_levels_FUNK.pdf. Published 10/19/2017.

Garcia A. Will 'Nebulized Hydrogen Peroxide' Help You Avoid Contracting COVID-19? Disinformation, Fact Checks Web site. www.truthorfiction.com/joseph-mercola-coronavirus-nebulized-hydrogen-peroxide/. Published 2020. Updated 4/14/2020. Accessed 11/12/2020.

Gautret P, Lagier J-C, Parola P, et al. Hydroxychloroquine and azithromycin as a treatment of COVID-19: Results of an open-label non-randomized clinical trial. Int J Antimicrob Agents. 2020;105949

Geleris J, Sun Y, Platt J, et al. Observational study of hydroxychloroquine in hospitalized patients with Covid-19. N Engl J Med. 2020;

Gleberzon B, Lameris M, Schmidt C, Ogrady J. On Vaccination & Chiropractic: When ideology, history, perception, politics and jurisprudence collide. J Canadian Chiropractic Assoc. 2013;57(3):205.

Godlee F, Smith J, Marcovitch H. Wakefield's article linking MMR vaccine and autism was fraudulent. Br Med J. 2011;342

Grabenstein JD. What the World's religions teach, applied to vaccines and immune globulins. Vaccine. 2013;31(16):2011–23.

Grein J, Ohmagari N, Shin D, et al. Compassionate use of remdesivir for patients with severe Covid-19. N Engl J Med. 2020;

Gupta R, Morain SR. Ethical allocation of future COVID-19 vaccines. J Med Ethics. 2020;47:3.

Hemilä H, Chalker E. Vitamin C for preventing and treating the common cold. Cochrane Database of Systematic Reviews. 2013. www.cochranelibrary.com/cdsr/doi/10.1002/14651858.CD000980.pub4/abstract. Published 1/31/2013.

Hinman A. Eradication of vaccine-preventable diseases. Annu Rev Public Health. 1999;20(1):211–29.

Hirschtick RE, Dyrda SE, Peterson LC. Death from an unconventional therapy for AIDS. Ann Intern Med. 1994;120(8):694.

History. Dr Jenner's House Web site. https://jennermuseum.com/history. Published 2020. Accessed 21 Oct 2020.

Huang WT, Chang S, Miller ER, et al. Safety assessment of recalled Haemophilus influenzae type b (Hib) conjugate vaccines – United States, 2007–2008. Pharmacoepidemiol Drug Saf. 2010;19(3):306–10.

Jarvis WT. Quackery: A national scandal. Clin Chem. 1992;38(8B Pt 2):1574–86.

Jonsen AR, Siegler M, Winslade WJ. Clinical ethics: A practical approach to ethical decisions in clinical medicine. Eighth edition. Ed. New York: McGraw-Hill Education; 2015.

Jordan K, Mackey D, Garvey E. A 39-year-old man with acute hemolytic crisis secondary to intravenous injection of hydrogen peroxide. J Emerg Nurs. 1991;17(1):8–10.

Kadkhoda K. Herd Immunity to COVID-19: Alluring and Elusive. American Journal of Clinical Pathology. 2021. https://academic.oup.com/ajcp/advance-article/doi/10.1093/ajcp/aqaa272/6063411. Published 1/5/2021.

Kim AH, Sparks JA, Liew JW, et al. A rush to judgment? Rapid reporting and dissemination of results and its consequences regarding the use of hydroxychloroquine for COVID-19. Ann Intern Med. 2020;

Kollaritsch H, Rendi-Wagner P. Principles of immunization. Travel Med Expert Consult-Online Print. 2012;67

Kukla R. Resituating the principle of equipoise: Justice and access to care in non-ideal conditions. Kennedy Inst Ethics J. 2007;17(3):171–202.

Kurian GT, Harahan JP. A historical guide to the U.S. government. New York: Oxford University Press; 1998.

Kuwabara N, Ching MSL. A review of factors affecting vaccine preventable disease in Japan. Hawaii J Med Pub Health J Asia Pacific Med Pub Health. 2014;73(12):376–81.

Luño AR. Ethical reflections on vaccines using cells from aborted fetuses. Nat Catholic Bioethics Quarterly. 2006;6(3):453–9.

Magagnoli J, Narendran S, Pereira F, et al. Outcomes of hydroxychloroquine usage in United States veterans hospitalized with Covid-19. Med1, 1–14, November 1, 2020. Elsevier, https://doi.org/10.1016/j.medj.2020.06.001.

Meikle J, Boseley S. MMR row doctor Andrew Wakefield struck off register. The Guardian. 2010. www.theguardian.com/society/2010/may/24/mmr-doctor-andrew-wakefield-struck-off. Published 5/24/2010.

Milazzo S, Lejeune S, Ernst E. Laetrile for cancer: A systematic review of the clinical evidence. Support Care Cancer. 2007;15(6):583–95.

Mill JS. Utilitarianism, Liberty, Representative Government. 1859. Sagwan Press edition published August 23, 2015.

Mitchell GW. A brief history of triage. Disaster Med Pub Health Preparedness. 2008;2(S1):S4–7.

MRC-5. ATCC Products Web site. www.atcc.org/Products/All/CCL-171.aspx?geo_country=us#characteristics. Published 2020. Accessed 11/28/2020.

Offit PA. Thimerosal and vaccines — A cautionary tale. N Engl J Med. 2007;357(13):1278–9.

Pelčić G, Karačić S, Mikirtichan GL, et al. Religious exception for vaccination or religious excuses for avoiding vaccination. Croatian Med J. 2016;57(5):516.

Pittler MH, Brown EM, Ernst E. Static magnets for reducing pain: Systematic review and meta-analysis of randomized trials. CMAJ. 2007;177(7):736–42.

Plotkin SA. Studies of immunization with living rubella virus. Trials in children with a strain cultured from an aborted foetus. Am J Dis Children. 1965;10:381–9.

Pruss AR. Complicity, fetal tissue, and vaccines. Nat Catholic Bioethics Quarterly. 2006;6(3):461–70.

Robbins R, Weiland N. A U.S. program aiming to channel a huge number of vaccines through pharmacies will start soon. New York Times. 2021. www.nytimes.com/live/2021/02/02/world/covid-19-coronavirus#covid-vaccine-pharmacies. Published 2/24/2021.

Ruijs WLM, Hautvast JLA, Kerrar S, Van der Velden K, Hulscher MEJL. The role of religious leaders in promoting acceptance of vaccination within a minority group: A qualitative study. BMC Public Health. 2013;13(1):1–8.

Saper RB, Phillips RS, Sehgal A, et al. Lead, mercury, and arsenic in US-and Indian-manufactured Ayurvedic medicines sold via the internet. JAMA. 2008;300(8):915–23.

Savaliya AA, Shah RP, Prasad B, Singh S. Screening of Indian aphrodisiac ayurvedic/herbal healthcare products for adulteration with sildenafil, tadalafil and/or vardenafil using LC/PDA and extracted ion LC–MS/TOF. J Pharm Biomed Anal. 2010;52(3):406–9.

Šeruga M, Grgić J, Mandić M. Aluminium content of soft drinks from aluminium cans. Z Lebensm Unters Forsch. 1994;198(4):313–6.

Slaoui M, Hepburn M. Developing safe and effective Covid vaccines—Operation warp Speed's strategy and approach. N Engl J Med. 2020;383(18):1701–3.

Smith K. Edward Jenner and the small pox vaccine. Front Immunol. 2011;2(21)

Spyropoulos AC, Ageno W, Albers GW, et al. Rivaroxaban for thromboprophylaxis after hospitalization for medical illness. N Engl J Med. 2018;379(12):1118–27.

The Dakar Declaration on Vaccines. World Health Organization. 2014. www.afro.who.int/publications/religious-leaders-declaration-vaccination-dakar. Published 3/25/2014.

The Eradication of Smallpox. Edward Jenner and the first and only eradication of a human infectious disease. Nat Med. 2001;7(1):15.

Thimerosal and Vaccines. U.S. Food and Drug Administration Web site. www.fda.gov/vaccines-blood-biologics/safety-availability-biologics/thimerosal-and-vaccines. Published 2018. Accessed 11/28/2020.

Thimerosal in Vaccines. Centers for Disease Control and Prevention. 2015. www.cdc.gov/vaccine-safety/concerns/thimerosal/index.html. Published 10/27/2015.

Tostanoski LH, Wegmann F, Martinot AJ, et al. Ad26 vaccine protects against SARS-CoV-2 severe clinical disease in hamsters. Nat Med. 2020;26(11):1694–700.

Triggle N. Wakefield and autism: the story that will not go away. BBC News 2010. http://news.bbc.co.uk/2/hi/health/8481583.stm. Published 1/28/2010.

Turner A. Jewish decisions about childhood vaccinations: The unification of medicine with religion. Paediatrics Health. 2017;5(1):1.

Turner RB, Bauer R, Woelkart K, Hulsey TC, Gangemi JD. An evaluation of Echinacea angustifolia in experimental rhinovirus infections. N Engl J Med. 2005;353(4):341–8.

Ulbricht C, Ko E. Common Complementary and Integrative Medicine Health Systems. In: Krinsky DL, ed. Handbook of Nonprescription Drugs: An Interactive Approach to Self-Care 19th Edition. American Pharmacists Association; 2017.

Ulbricht C, Cohen L, Lee R. Complementary, alternative, and integrative therapies in cancer care. In: DeVita V, Lawrence T, Rosenberg S, eds. Cancer: Principles & Practice of Oncology, 11th Ed. 2018.

Vaccine Excipient Summary. Centers for Disease Control and Prevention Web site. www.cdc.gov/vaccines/pubs/pinkbook/downloads/appendices/b/excipient-table-2.pdf. Published 2020. Accessed 11/18/2020.

Vaccine Recalls. Center for Disease Control and Prevention Web site. www.cdc.gov/vaccinesafety/concerns/recalls.html. Published 2020. Accessed 11/7/2020.

Vaccines & Immunizations. Centers for Disease Control and Prevention Web site. www.cdc.gov/vaccines/vpd/vaccines-list.html. Published 2018. Accessed 11/21/2020.

Ventilator Allocation Guidelines. New York State Task Force. 2015. www.health.ny.gov/regulations/task_force/reports_publications/docs/ventilator_guidelines.pdf.

Wakefield AJMS, Anthony A, et al. Ileal-lymphoid-nodular hyperplasia, non-specific colitis, and pervasive developmental disorder in children. Lancet. 1998;351:637–41.

Wakefield A, Murch S, Anthony A. Ileal-lymphoid-nodular hyperplasia, non-specific colitis, and pervasive developmental disorder in children (Retraction of vol 351, pg 637, 1998). Lancet. 2010;375(9713):445.

Walker BF, French SD, Grant W, Green S. Combined chiropractic interventions for low-back pain. Cochrane Database Syst Rev. 2010;4. www.cochranelibrary.com/cdsr/doi/10.1002/14651858.CD005427.pub2/epdf/full

Wang S-M, Kain ZN, White P. Acupuncture analgesia: I. the scientific basis. Anesth Analg. 2008;106(2):602–10.

White DB, Lo B. A framework for rationing ventilators and critical care beds during the COVID-19 pandemic. JAMA. 2020; https://jamanetwork.com/journals/jama/fullarticle/2763953

White DB, Lo B. Mitigating inequities and saving lives with ICU triage during the COVID-19 pandemic. Am J Respir Crit Care Med. 2021;203(3):287–95.

WI-38. ATCC Products Web site. www.atcc.org/Products/All/CCL-75.aspx?geo_country=us#characteristics. Published 2020. Accessed 11/28/2020.

Wong A. The ethics of HEK 293. Nat Cathol Bioeth Q. 2006;6(3):473–95.

Chapter 13
Case Studies and Policy Scenarios

So far, this text has presented the foundational principles of pharmacy ethics, followed by clinical applications in a variety of domains: reproductive ethics, end-of-life, conscience claims, vaccines, pandemics, and community pharmacies. It is now time for you, the reader, to begin integrating these concepts into your own clinical and academic work, to help you navigate ethical practice in your day-to-day professional life.

This chapter will present a series of case studies for you to analyze. Each one will give enough background information to help you get started. However, it may also require a review of ethical principles and a literature review to determine best practices in similar scenarios. The first set of cases will focus on specific clinical situations; we will then present some broader, more policy-based examples. Though most will be very practical and realistic, a few will be more theoretical and will serve as a review of basic clinical ethics. Note: to make these cases maximally beneficial, some of them are not specific to pharmacists. However, we have also included many cases that directly relate to pharmacy practice.

Clinical Ethics Analysis: Key Steps

The following set of questions are intended as a guide[1] to help you with ethical reasoning. Not all of these points will apply in every case. The goal is to arrive at moral conclusions and recommendations that resolve the issues logically and accurately.

---

[1] The methodology presented here above is a shortened summary of the "Four Topics" method, sometimes called the "Four Boxes." A complete explanation of this approach can be found in: Jonsen AR, Siegler M, Winslade WJ. *Clinical ethics: a practical approach to ethical decisions in clinical medicine.* Eighth edition. ed. New York: McGraw-Hill Education; 2015, pp. 3–9.

What Are the Medical Facts?

We should begin our analysis of the medical problem. This step may include some of the following types of questions: What is the diagnosis? Is it acute or chronic, treatable, or terminal? What other clinical issues complicate the picture? What is the goal of treatment, and can it be achieved? In short, can the current treatment plan help the patient (beneficence) and avoid harm (non-maleficence)?

What Is the Central Ethical Question?

The key question may be obvious and expressly stated. Here are a few examples:
- Can clinicians override this patient's refusal to receive treatment?
- Is it ethically permissible to discontinue ventilator management in this patient with end-stage lung failure?
- Is it ethically appropriate to dispense insulin to this diabetic patient who cannot get a prescription from his physician during a holiday?

Often, in a hospital or clinic setting, clinicians or family members may ask for an ethics consultation because of a conflict. For example: "Family members are fighting over the management of their elderly mother in a coma. Help us figure out the right thing to do." In such a case, it will be up to the consultant to come up with the central ethical question. Nonetheless, clarity about the central question will avoid confusion later on.

What Are the Legal and Situational Constraints?

These are the contextual features that differentiate between what one might wish to do versus what can actually be done. The desired intervention may be illegal or outside of standard medical practice. Some constraints may be institutional, such as the inability to perform an abortion in a Catholic hospital. Some may be situational, such that specific interventions are not currently available (e.g., an acute shortage of ICU beds).

Who Are the Stakeholders, and Who Decides?

It is always essential to respect a patient's autonomy. But does the patient have decision-making capacity? If not, who may legally and ethically act as a surrogate decision-maker? If no advance directive is in place, who makes decisions when a conflict arises among two or more potential surrogates?

What Is the Patient's Quality of Life?

As we discussed in Chap. 9 on end-of-life ethics, the judgment of a patient's quality of life is best made by the one living that life or by the patient's family, not clinicians. Nonetheless, quality of life considerations may significantly influence clinical choices. Distributive justice requires that clinicians strive to protect vulnerable patients who may not be able to express their desires adequately.

What Religious, Moral, or Cultural Factors Influence the Outcome?

Some limitations may arise because of a patient's or family's religious affiliation, e.g., impermissibility of blood transfusions for an observant Jehovah's Witness. Similarly, should prayer, faith-healing, or alternative therapies play a role? When should certain clinically-neutral cultural practices be permitted, and when might these interfere with treatment goals?

What Ethical Principles Have a Bearing on the Case?

After considering all of the above factors, it is essential to bring ethical theory to bear on the central question. This analysis begins with medical principlism (beneficence, non-maleficence, distributive justice, and autonomy), but other ethical theories will be worth considering. For example, maximizing the clinical outcome always involves prudential reasoning from a utilitarian framework. Employing strong communication skills to resolve an impasse among stakeholders requires virtue ethics, where the clinician employs compassion and empathy to help resolve the conflict. Religious thinking is essentially deontological, based on principles, and the clinician who respects the patient's spiritual values understands divine-command ethics. As a final example, determining the best practices for one's profession (e.g., pharmacy practice) can be very Kantian, inasmuch that the clinician strives to make a decision that could be a model (or "maxim") for the profession as a whole (see Chap. 2 to review these theories).

Based on these Principles, What Ethical Duties Conflict, and What Is Their Best Resolution?

The preliminary case analysis will generate specific ethical duties, not all of which can be fulfilled entirely within a complex clinical scenario. The duties must be then arranged *hierarchically*; in other words, it is essential to determine which ones should take precedence in this case.

What Is the Ethical Conclusion?

Ethical conclusions should be worded in the following form: "It is ethically permissible to…" or "It is ethically problematic to…" Here are some possible examples:

- In this elderly stroke victim, it is ethically permissible to discontinue the ventilator based on his advance directive, even if the family disagrees.
- In this 20-year-old woman in hemorrhagic shock, it is ethically problematic to give her blood transfusions, in light of her devout Jehovah's Witness faith.

The case analysis should also include additional recommendations from an ethics perspective, which may provide alternative plans of action or supplemental advice for carrying out the ethical dictates.

Key Steps in a Clinical Ethics Analysis:

1. What are the medical facts?
2. What is the central ethical question?
3. What are the legal and situational constraints?
4. Who are the stakeholders, and who decides?
5. What is the patient's quality of life?
6. What religious, moral, or cultural factors influence the outcome?
7. What ethical principles have a bearing on the case?
8. Based on these principles, what ethical duties conflict, and what is their best resolution?
9. What is the ethical conclusion?

Clinical Case Studies

In the following cases and scenarios, the first two will provide a complete ethical evaluation by way of demonstration. The format will follow that of a formal ethics consultation, including the key ethics question, a case summary, an ethical analysis, recommendations, and suggested academic references. All of the remaining cases will be left for the reader to analyze. We will include discussion questions and occasionally some resources to help guide your thinking.

Case Study #1: Dealing with a Difficult Patient

Jerry S. is a 65-year-old retired business executive with chronic high blood pressure and type 2 diabetes. He is obese (5' 11", 325 lbs., BMI 45.3 kg/m^2) and has a sedentary lifestyle.

Though retired, Jerry still does various consulting jobs, with irregular work hours. He is divorced and lives alone. He has resisted several attempts by his physician to get him to curb his two pack per day smoking habit, to exercise, or to eat a healthier diet.

Two days ago, Jerry complained of heart palpitations during a business meeting, then suddenly passed out, and his associates called 911. He was admitted to the Coronary Care Unit (CCU) in acute atrial fibrillation, with low systolic blood pressure. After ruling out myocardial infarction in the CCU, he was anticoagulated on enoxaparin (Lovenox®) and given medication to slow his heart rate. The cardiologist wants to schedule cardioversion to restore him to sinus rhythm, but the patient refuses.

You are the ethics consultant for the hospital. The cardiologist wants to discharge the patient for being so uncooperative and for refusing to take his meds or modify his lifestyle. The patient's ex-wife and the primary care physician are trying to see that he gets the proper care he needs. The social worker has called for an ethics consult: "Please come and review the chart, talk with the patient, his ex-wife, and the caregivers, and help us sort all this out."

Additional Medical Facts

- Type 2 diabetes, poorly controlled
- Poorly adherent with diet and medication recommendations

Medications

- metformin (Glucophage®), 1000 mg twice daily
- losartan (Cozaar®), 50 mg daily
- atorvastatin (Lipitor®), 20 mg daily
- enoxaparin (Lovenox®), 1 mg/kg subcutaneously every 12 hours

Family and Lifestyle Issues
Two male children are both married and live away from their father. The ex-wife has a reasonable relationship with Jerry and is willing to help. Jerry watched his older brother deal with complications of his type 2 diabetes, including losing a limb to amputation from gangrene. So Jerry figures that the condition is genetic, and nothing he can do will change things. He has been chronically angry and frustrated after his failed marriage and simply wants to be left alone. He hates being in the hospital.

Key Ethics Question Is it ethically permissible to discontinue clinical care in this uncooperative patient with atrial fibrillation?

Case Summary

Jerry is a 65-year-old retired business executive with chronic high blood pressure and type 2 diabetes. He is obese, a heavy smoker, and has a sedentary lifestyle. He was admitted at this time with acute atrial fibrillation. After ruling out an MI and placing him on anticoagulation, the cardiologist has recommended cardioversion, which the patient refuses. The cardiologist is considering discharging the patient for his lack of cooperation.

Jerry's emotional health is complicated by chronic anxiety, depression, and anger after a failed marriage. His poor health and his experience with a brother's adverse medical outcome have made him suspicious about hospital care and fatalistic about his prospects. His ex-wife is concerned and is available to help.

Ethics Discussion

In this case, there is a conflict of goals between an emotionally overwrought patient and his treating physician. At risk is the therapeutic relationship that allows for expeditious management of his cardiac condition and eventual discharge. Currently, there are four major problems to be addressed:

1. Jerry has been admitted with acute atrial fibrillation, which remains untreated. This problem puts him at risk for stroke and further hypotensive episodes with syncope.
2. The patient's lack of cooperation has interfered with a treatment approach consistent with standards of care.
3. Jerry may have other mental health issues that remain underdiagnosed and untreated.
4. The patient's combative attitude makes it harder for busy CCU health care team members to stay focused on his needs.

Patient autonomy, the overarching principle of most healthcare interactions, is not absolute. Though it appears that Jerry is making up his own mind, his decision-making capacity is in doubt. If he is in an acute depressive state, that is all the more reason to question his judgment.

It is certainly not appropriate for the cardiologist to have to deal with this patient's mental health issues. That will require psychiatric consultation. In the meantime, the social worker or chaplain should try to get more information to clarify the reasons for Jerry's fears and lack of cooperation. Jerry's ex-wife is the one family member who remains involved and engaged and should be considered an ally in assessing Jerry's ability to cooperate with his treatment.

At the same time, it is not appropriate to discharge the patient until his acute medical condition has stabilized as much as possible. Physician frustration does not justify summarily discharging the patient, though it may be appropriate to involve another cardiology colleague.

Ultimately, the autonomous decisions of a patient with decision-making capacity will have to be honored. In this case, Jerry is not refusing every treatment; he is merely refusing cardioversion. It will remain a clinical matter whether any alternative clinical strategies exist. If not, then the patient must be carefully and accurately told of the risks of forgoing the recommended clinical approach before sending him home.

Recommendations

1. Discharging the patient from the coronary care unit is ethically premature at present, before exploring all available and alternative treatment strategies.
2. Attempts should be made by social services, mental health professionals, and the clinical team to obtain Jerry's consent to cardioversion as the single best treatment for his atrial fibrillation.
3. Even if he eventually consents to the cardiologist's advice, Jerry has several outstanding and unresolved mental health and behavioral issues. He will need careful follow-up and help to be adherent with medications for his cardiac condition, diabetes, and hypertension.
4. If Jerry remains resistant to further medical interventions, then it is not unethical to discharge him, as long as attempts continue to treat him medically as an outpatient.

Suggested Reference Guidelines for Difficult Provider-Patient Relationships. *Midwest Bioethics Center* (see reference).[2]

Case Study #2: Dealing with Cultural Beliefs

Anjali M. is a 72-year-old woman, an immigrant from India, brought to the office of an internist with a busy academic practice in a teaching hospital. The physician and internal medicine resident together interview Anjali, who is accompanied by her husband. In halting English, Anjali complains of poor appetite and moderate weight loss, which she has noted for the past 6 weeks or so. Her husband, Vihaan, speaks much better English. He adds that his wife has also appeared more tired lately.

With the attending physician's supervision, the resident performs a thorough physical exam. During the exam, he asks Vihaan to step out of the room, which he initially refuses to do, but then relents. Upon deep abdominal palpation, the resident finds a large, firm mass in the epigastric region. The mass is immobile and nontender. The resident asks his attending to examine the patient as well, and the mentor confirms the exam findings. The rest of the exam is unremarkable.

Vihaan is brought back into the room, and the attending physician starts to explain his concerns to Anjali. He begins to tell her that there is a mass in her abdomen that will require further tests, but Vihaan interrupts. He demands to speak to the doctors privately, with Anjali out of the room. The physicians reluctantly agree.

Vihaan tells them, "You may not discuss the treatment of my wife with her – you can only discuss this with me. In our culture, the husband is in charge, and the wife must obey. I will decide what is best for her." The attending internist explains that

---

[2] Guidelines for Providing Ethical Care in Difficult Provider-Patient Relationships. *Midwest Bioethics Center.* 2000. www.practicalbioethics.org/files/members/documents/Supplement_16_3.pdf

Anjali needs a CT scan and blood work to confirm the diagnosis of pancreatic malignancy so that she can begin treatment.

Vihaan tells the doctors that he wants a second opinion from his community. He wishes to consult with an Ayurvedic practitioner, to have him use vibhuti (holy ash) on her body. He is sure that this will heal her and knows that her illness was a result of bad karma from several instances when she disobeyed him last year. He is not willing for her to have any further tests until this is done. The internist and his resident are at a loss. They turn to you, the hospital pharmacist, to help sort all this out, knowing of your expertise in alternative therapies.

After obtaining permission from the couple, you have a phone conversation with Dr. Chandranash Pradesh, an Ayurvedic healer. He says, "Thank you for consulting with me. I do not often get to compare notes with scientifically-trained doctors. Vibhuti is sacred ash made from the burning of plant material and is a carrier or transport device for sacred energy or vibration.[3] Because such diseases as you describe have a clear spiritual cause, vibhuti ash must be applied first. However, once the karma has been adjusted, there is no reason not to use additional treatments. The patient may begin chemotherapy after I finish the rite of healing."

Key Ethics Question How can the healthcare team honor autonomy, beneficence, and Hindu cultural beliefs in this patient with an advanced abdominal malignancy?

Case Summary

Anjali M. is a 72-year-old immigrant woman from India, accompanied by her husband. Symptoms and physical exam findings point towards the possibility of a severe pancreatic malignancy, which would typically elicit a series of confirmatory and staging medical tests.

The husband has a strong desire to be the sole decision-maker is his wife's care. He is also strongly influenced by Ayurvedic practices and the belief that his wife's illness may be the result of bad karma, perhaps related to perceived marital failings on her part. He will not permit further diagnostic or therapeutic interventions until consultation with an Ayurvedic practitioner and the ritual use of vibhuti (holy ash) on her body. The patient is willing to go along with all this. For his part, the Ayurvedic practitioner is not putting up any barriers to diagnostic tests or chemotherapy after he completes his ritual.

Ethics Discussion

Many alternative medical practices have not undergone the rigors of randomized clinical trials, FDA approval, or other aspects of evidence-based medicine. Yet these are often an essential part of the cultural context for patients. Though many of us may be skeptical of relying on unconventional therapies, we should not allow our doubts to interfere with the therapeutic covenant we hold with our patients.

---

[3] Vibhuti - How and Where Should We Apply It. Isha Foundation Web site. https://isha.sadhguru.org/us/en/wisdom/article/why-we-do-what-we-do-the-science-of-vibhuti. Published 2019. Accessed 10/4/2020.

Holy ash is a recognized treatment modality within Ayurvedic medicine, which has a long and rich tradition. There is no evidence in this case that the desire of the patient and her husband to use this method will directly harm her in any way. It is even conceivable that the patient may exhibit a more robust response to conventional therapy, having met her perceived spiritual need for ritual cleansing. While it may not be convenient for scheduling, there is no evidence that participation in this benign ritual will substantially delay her eventual diagnosis and treatment.

For a patient, or in this case, a surrogate, to exercise autonomy, they must be able to affirm the validity of any resource they choose. While we may disagree with a particular decision, disagreement itself does not make the decision irrational. In this case, relying on both Ayurvedic and modern medicine is culturally rational.

Of more significant concern is the matter of proper informed consent, given the husband's exclusivism and desire to make all of the significant decisions. Nonetheless, from the information we have obtained, it appears that Anjali agrees with this arrangement. Her husband is, therefore, an appropriate surrogate for the time being.

Recommendations

1. It is ethically permissible to slightly delay further diagnostic procedures in this patient until she has completed a benign Ayurvedic ritual.
2. Further diagnostic and therapeutic interventions should proceed quickly, however. The husband may serve as the principal decision-maker, as long as the clinical team feels that he is acting in his wife's best interests.
3. The clinical team should be careful not to undermine the husband, nor to express skepticism about his cultural and spiritual beliefs. Upon completion of diagnostic testing, however, the results should only be shared with the husband with his wife present.
4. Nonetheless, cultural sensitivity is not absolute. There is great potential for abuse of informed consent with an overly-dominant husband, and subtle recrimination due to perceived failings on the wife's part. If these conflicts occur, the clinical team should consider consulting again with the Ayurvedic practitioner. He is undoubtedly a therapeutic ally and might serve in a similar role as a hospital chaplain.

Suggested Reference Cross-Cultural Issues and Diverse Beliefs. University of Washington Department of Bioethics and Humanities (see reference).[4]

---

[4] Cross-Cultural Issues and Diverse Beliefs. University of Washington Department of Bioethics and Humanities Web site. https://depts.washington.edu/bhdept/ethics-medicine/bioethics-topics/detail/59. Published 2018. Accessed 10/4/2020.

Case Study #3: The Ethics of Palliative Care[5]

Gloria J. is a 48-year-old woman with metastatic cancer of the breast. Her disease started in the left breast, and she underwent a modified radical mastectomy (removal of the left breast, with a sampling of lymph nodes for staging). She also had chemotherapy and local radiation to the left chest wall.

Unfortunately, after 3 years, Gloria's cancer has returned. Now it has metastasized to her ribs and spine, and she is having increasingly severe bone pain. Further chemotherapy would offer a 30% chance of improvement, with many side effects.

You are a pharmacist working with the palliative care team, and Gloria asks for your advice.

Questions for Discussion

1. Would you recommend that she undergo further chemotherapy? Which medical ethics principles might help to inform her decision?
2. Gloria is in severe pain. How should she cope with this? Her physicians are willing to prescribe high doses of narcotics, but these make her sleep all the time. When she wakes up, her pain has returned. What should she do? Is palliative sedation (sometimes called terminal sedation) ethically acceptable? Is this equivalent to euthanasia?
3. One of the physicians is worried that too high a dose of narcotics will depress the patient's respiratory drive. As Gloria gets weaker, he is concerned that the drugs may shorten her life. Can you ethically justify increasing doses of narcotics in this setting? If so, on what basis?

Suggested Reference "ASHP Guidelines on the Pharmacist's Role in Palliative and Hospice Care" (see reference) (Herndon et al. 2016).

Case Study #4: End-of-Life Decision Making

Steve C. is a 63-year-old engineer living in Grand Rapids, Michigan. After a vigorous professional life, he has been looking forward to retirement. Just 6 weeks ago, he began to notice that his clothes were fitting more loosely than usual and that he was losing weight. His appetite has decreased, and he has a growing ache in his lower back.

Steve went to see his family physician, who was worried enough to send him to a specialist. Now, after 2½ weeks of testing, Steve knows the truth. An abdominal CT scan revealed a mass in his pancreas. A radiologist passed a needle into the mass with ultrasound guidance, removing tissue for biopsy. The pathologist's report arrived yesterday, and the diagnosis is an incurable malignancy of the pancreas.

---

[5] Note that this case is essentially the same as that of John Walters (Case #1) from Chap. 1.

Steve has always been an active man, and the prospect of chemotherapy with all its complications and side effects does not appeal to him. Besides, he is a pragmatist and knows when it is his time to go. He does not wish to suffer the severe pain that his current diagnosis offers. Steve's wife left him five years ago. His two daughters are grown, with active lives of their own. They have never been close to their father. Steve has always been a proud, private person, and there are very few people in whom he will confide. You are his golf partner and his only real friend. Since you are also a community pharmacist, Steve has often relied on your informal advice in the past. Just this morning, Steve stops by your house to ask what you think of him seeking to end his life.

Questions for Discussion

1. If Steve were a resident of certain states (which states?), he could legally obtain a prescription for lethal drugs to end his life, provided he meets certain conditions. What are those requirements?
2. What trends in our society have contributed to the current change in attitudes towards medically-assisted dying?
3. You are Steve's best friend. What advice will you give him? From the perspective of ethical theory and medical principlism, provide a moral rationale both for and against using assisted suicide in this case. Be sure to use arguments he can readily understand.

Case Study #5: Abortion and Malignancy

Angela W. is a 34-year-old mother of two young boys, ages 2 and 4 years. Angie used to work part-time as a clinic nurse, but she recently gave up her nursing job when she found out she was pregnant with her third child. Angie's husband, Brad, is 36 years old. He has worked as an engineer with a local manufacturing firm for the past 8 years. Brad and Angie are both very excited about raising their family, since they have struggled in the past with infertility issues and have had two miscarriages. Both are religious and attend a small Lutheran church.

Recently, Angie became pregnant once again. She returned home from her first prenatal visit with troubling news. Angie is indeed 9 weeks pregnant, confirmed by ultrasound. However, a routine exam by the doctor also revealed a hard, non-tender, 2 cm. mass in the upper outer quadrant of Angie's left breast. Some smaller hard masses are palpable in the left axilla.

A lumpectomy and lymph node biopsy confirmed the diagnosis of poorly differentiated infiltrating ductal carcinoma of the breast, with positive lymph nodes. An oncologist consulting on the case has diagnosed Angie with a Stage II breast malignancy, which is estrogen receptor positive and progesterone receptor negative (ER+/PR-). A chest x-ray and various scans show no evidence of metastatic spread.

The oncologist has recommended termination of Angie's pregnancy, with immediate radiation therapy to the left chest wall and axilla, followed by an aggressive course of chemotherapy and hormonal treatment with the estrogen-receptor blocking drug tamoxifen. She assures Angie and Brad that this plan gives her a better than 80% chance of surviving more than five years.

Angie and Brad are devastated by the news and have asked for a few days to think things over. They made an appointment with Reverend Barry, the pastor of their church. His wife, Susan, works as an inpatient pharmacist at the local community hospital.

How should Rev. Barry and Susan advise Angie and Brad?

Questions for Discussion

1. Write a summary of the medical indications for the advice given. Even though there is no current evidence of metastases, what factors point to a more aggressive disease course in this patient? How does pregnancy complicate this picture? Explain the apparent urgency on the part of the oncologist.
2. Using arguments from ethical theory and medical principlism, how would you make a case for Angie to refuse radiation, chemotherapy, and hormonal therapy at this time, to preserve the life of her unborn child? Note: the Roman Catholic moral tradition will likely be an important influence for members of a Lutheran church. If Angie declines these treatments for now, is such a decision morally required by divine command ethics, or is it heroic?
3. Rev. Barry may be a devout Lutheran pastor, but he is also very pragmatic. Using arguments from ethical theory and medical principlism, how would you make a case for Angie to have an abortion, followed by radiation, chemotherapy, and hormonal therapy?

Suggested Reference Cordeiro CN, Gemignani ML. Breast Cancer in Pregnancy: Avoiding Fetal Harm When Maternal Treatment Is Necessary. (Cordeiro and Gemignani 2017).

Case Study #6: Neglect in an 8-Year-Old Boy

Jonathan W. is an 8-year-old boy with a history of congenital cytomegalovirus infection that has left him non-verbal, with severe cognitive disabilities. He also has esophageal reflux, which complicates his ability to get adequate nutrition. The pediatric service is evaluating the child and his mother because the boy has failed to thrive. His weight is less than the third percentile for his age, with no measurable weight gain for the past 3 years, and recent weight loss. Recent visits with the pediatrician have shown the child to be less responsive and not smiling.

The child's father also lives in the home, along with another sibling, age seven, without any medical issues. When asked about the child's eating habits, the mother becomes somewhat defensive. She claims that Jonathan eats just fine and that he is maintaining his weight.

The pediatric surgeon has recommended a Nissen fundoplication procedure to correct the child's reflux. However, the gastroenterologist feels that the boy is not nutritionally ready for such major surgery. He has recommended that the placement of a gastrostomy tube to supplement feedings will help the child to gain weight. The mother, however, is hesitant to agree to the procedure, because she feels that doing so would mean that she is a failure as a mother. During the past 3 months, she has failed to keep her appointments with the healthcare team and has steadfastly refused to consider the gastrostomy procedure.

The pediatric care coordinator has called you, the ethics consultant, to give your input. She says, "We don't know what to do. Jonathan is losing ground and needs nutritional intervention. We need to figure out how to get her mother to cooperate."

Questions for Discussion

1. What is the central ethics question in this case?
2. How could you discuss with the mother the reasons for her refusal?
3. Is it ethically and legally possible to override the mother's wishes, to meet the nutritional needs of her child?

Suggested References

- Look up "justified hard paternalism" in *Principles of Biomedical Ethics (eighth Ed)*. (Beauchamp and Childress 2019)
- Woods M. Overriding Parental Decision to Withhold Treatment. (Woods 2003)

Case Study #7: Helping a Patient with Substance Abuse Disorder[6]

It is Christmas day, and medical offices are closed, but your community pharmacy is open. You are on duty as a pharmacy intern. Jeffrey M. is a 36-year-old man who approaches the counter to pick up his monthly supply of buprenorphine. He has been stable on medication-assisted treatment (MAT) for opioid addiction for the past 6 months. He has a steady job and was able to move back in with his wife and two children, whose photo he carries with him at all times. He is completely out of medication and shows visible signs of distress.

This provider is known to wait until patients are entirely out of their meds before releasing a new prescription order. Unfortunately, with the holiday, he failed to provide the new order for buprenorphine. Jeffrey has not even had his morning dose.

You attempt to contact the provider, but the office is closed, and no one is listed as on-call. Desperate, you contact the nearest three recovery centers and the emergency mental health clinic. All of them have policies that withhold MAT for new patients until 3 days after admission. You even contact the emergency department of the local hospital. They inform you they are unable to provide MAT or withdrawal-mitigating therapies unless they have written documentation that the patient will

---

[6]This case study is taken from an actual clinical encounter, contributed by PharmD candidate Joshua Pearson (class of 2021).

transition to a recovery specialist. As a last resort, you ask your pharmacist to authorize a one-day emergency fill as permitted by state law. The pharmacist refuses, muttering something about addicts being liars.

You attempt to explain the situation to Jeffrey. His hands are visibly shaking and he admits he does not feel safe. He pulls out the photo of his children and looks at it, and then he says that the whole world is against him. He glances over at the counter where the buprenorphine stock bottle is visible and remarks, "So, there's no way in the universe that I can leave here with the medication I've been taking for months? There's no way that someone, *anyone*, would possibly loan me or give me the tablets? No, I imagine that would cost you your job."

Jeffrey's jaw tightens as he sighs and mutters, "I know where to get what I need." Frantic, you plead with him to wait – you know that the local dealers supply fentanyl-laced heroin, and there has been a recent spike in overdose deaths. You have seconds to offer Jeffrey some solution as you watch him head towards the door.

Questions for Discussion

1. As a pharmacy intern, is there anything you can do to help Jeffrey?
2. What are the legal constraints?
3. How should you, as an intern, respond to the pharmacist's behavior?
4. This unfortunate case may have negative repercussions for the patient, but also the clinician. What is the definition of *moral distress*, and how can clinicians alleviate it?

Suggested References

- McElrath K. Medication-Assisted Treatment for Opioid Addiction in the United States: Critique and Commentary (McElrath 2018)
- Lamiani G, Borghi L, Argentero P. When Healthcare Professionals Cannot Do the Right Thing: A Systematic Review of Moral Distress and its Correlates. (Lamiani et al. 2017)

Case Study #8: A Patient with a Massive Head Injury

Part I

Ralph W. is a 52-year-old man, thrown from his vehicle after a head-on collision on the interstate. He was transported by ambulance to the Emergency Department (ED) of a large community hospital with an apparent severe head injury. On arrival, he was urgently intubated and stabilized on a ventilator. Currently, he has the following physical findings:

- Visible penetrating skull fracture in the frontal region
- Pupils sluggishly reactive to light
- Deep tendon reflexes diminished, but present
- No reaction to painful stimuli
- No other apparent bodily injuries

The ED physician calls for an urgent surgical consult. The third-year resident on call arrives between surgery cases and does a cursory exam of the patient. After she notes that the patient does not react to painful stimuli, she declares the patient "brain dead" and orders all treatment efforts to cease and to turn off the ventilator. She then returns to the operating room.

The ED physician is not comfortable with these conclusions and calls you, the hospital ethicist, to consult on the case.

Questions for Discussion

1. Do you agree with the surgical resident's opinion?
2. What are the criteria for the clinical determination of brain death? Does the patient meet these criteria at present?
3. If the patient does meet the criteria for brain death, what additional steps are necessary before ventilator withdrawal might be warranted?

Part II

You and the ED physician call in a neurosurgeon to evaluate Ralph. After his exam and evaluation, he orders the transfer of the patient to the Intensive Care Unit (ICU), where he undergoes intensive management of his head injury and severe cerebral edema. You continue to follow the case as the ethics consultant.

Ralph's condition begins to stabilize, and his neurological exam improves. Three weeks after admission, the patient is breathing on his own without a ventilator. Currently, he has the following physical findings:

- He has deep tendon reflexes, and his pupils react to light.
- His heart rate and blood pressure are stable. His eyes are open, and he has normal sleep and wake cycles.
- Other than simple reflexes, however, he remains completely unresponsive to external stimuli. His eye movements are not purposeful and do not track objects in the room.

The patient is no longer receiving intravenous fluids. However, he has a surgically-placed feeding tube through his abdominal wall to maintain nutrition with tube feedings. The current diagnosis is a persistent vegetative state (PVS).

Questions for Discussion

1. How does PVS differ from brain death?
2. Does this patient have decision-making capacity? If not, who should make clinical decisions for him?
3. Family members report that Ralph would never wish to continue treatments just to prolong life without improving his quality of life. Is it appropriate to discontinue the feeding tube? Why or why not? Justify your answer using a proportionality argument.
4. In a large hospital, open ICU beds are always in short supply. Besides, such care is expensive, with daily costs in the thousands of dollars. The hospital administrator is putting a lot of pressure on you to discontinue the feeding tube and move the patient out. How should you respond? What ethical principle applies here?

Case Study #9: Should the Pharmacist Dispense this Medication?[7]

Late on a Sunday afternoon, a pharmacist overhears an escalating conversation between a nurse and Charles T., a patient in a small, rural hospital. Charles had been admitted for a transient ischemic attack and was now finally ready to go home. During his stay, several new medications were initiated, which were critical for him to continue taking at home. The discharge orders were ready, but the physician had failed to write prescriptions, leaving Charles unable to continue his therapy once released. Also, it is now late on Sunday, and there are no local pharmacies open to fill the orders even if they were provided.

The pharmacist approaches Charles and the nurse to investigate the situation and offer his help. The nurse provides the background information and tells the pharmacist that she was unsuccessful in reaching the physician. Charles describes his concern for not being able to continue his medications at home without prescriptions or an open pharmacy. Understanding the problem, the pharmacist asks to look at the discharge orders. Upon review, the physician's intent for the post-discharge medications is clear; all of the necessary details for valid prescriptions are present.

Questions for Discussion

1. Is it ethically permissible for the pharmacist to dispense medication without a prescription in this urgent situation?
2. What are the legal barriers?
3. What is the pharmacist's fiduciary duty, assuming that he decides to help in this situation?
4. Is there a way for the pharmacist to help Charles while mitigating any potential misunderstanding?

Case Study #10: Should This Patient Have Emergency Heart Surgery?[8]

Sister Caroline B. is an 84-year-old Catholic nun, being evaluated for heart disease. She has a history of angina, and a cardiac stress test showed significant coronary artery narrowing. Today, she was admitted to the cardiac catheterization lab for an angioplasty to dilate the narrowed vessel. She signed the consent form, which

---

[7] This case study first appeared in: Ward JA, Sullivan DM, Yates FD. A Pharmacist's Dilemma. *Ethics & Medicine*. 2019;35(1):13–14. Used with permission.

[8] This case was modified from one originally written by Dr. Robert Orr. Our analysis of it first appeared in Sullivan DM, Salladay SA. Is It Permissible to Forgo Emergent Restorative Surgery in This Case? *Ethics & Medicine*. 2009;25(1):17. Used with permission. The case (#3.10) was also later published in Orr RD. *Medical Ethics and the Faith Factor: A Handbook for Clergy and Health-Care Professionals*. Grand Rapids, Mich.: William B. Eerdmans Pub. Co.; 2009.

included standard wording to permit any other procedures required due to unexpected findings. She was told that the risks of today's procedure included emergency surgery, and she consented if such an operation became necessary to restore her health.

The angioplasty procedure was done about 2 hours ago. Immediately after the procedure, the patient's blood pressure dropped significantly. On return to the cath lab, the cardiologist found that the coronary artery was ruptured. A balloon pump was placed to sustain her vital circulation, and she was immediately transferred for emergency surgery to repair the damage. She is at this moment on the operating table, and the team is ready to begin the operation.

During preparation for surgery, Sister Caroline had new-onset generalized seizures, suggesting brain damage from lack of oxygen. Though survival of such emergency cardiac surgery is common, intact survival in an elderly patient with signs of brain damage from inadequate oxygen is questionable. There is a reasonable chance she could survive the crisis, but she will likely have some neurological impairment.

The cardiologist has just discussed the case with Sister Caroline's Mother Superior. The Mother Superior told him that Sister Caroline has always said she would not want resuscitation if she should become neurologically incapacitated. The Mother Superior confirmed this with a phone call to another nun who also knows Sister Caroline well.

The cardiac surgeon has just asked the cardiologist if he should go ahead with the operation. Because the cardiologist is hesitating, they both have asked for an emergency consult from you, the clinical ethicist.

The patient is on the operating table. What do you advise?

Questions for Discussion

1. All of this is an unexpected complication of a routine angioplasty. Is it likely that the patient will be restored to her original status before the elective procedure?
2. Who has the right and the responsibility to make treatment decisions for Sister Caroline? In her present condition, she lacks decision-making capacity, and the healthcare team must rely on a surrogate. Is her Mother Superior an appropriate surrogate?
3. Even though the patient is currently on the operating table, there is still time to reverse course. Either course of action (continuing or not continuing with surgery) is morally permissible, but which is *best*?

Case Study #11: Medical Marijuana for Chronic Pain[9]

Dr. Jeffrey Rogers works as a pharmacist in a collaborative practice with three physicians. Their clinic deals with the management of chronic pain disorders. Julia S. suffers from the symptoms of fibromyalgia. Though Dr. Rogers and the team have tried different approaches, none of them has eased her pain, and she has

---

[9] Note that this is Case #3 from Chap. 1.

become depressed. A friend told Julia how much the regular use of marijuana had helped her, so she is anxious to try it. Dr. Rogers realizes that no high-quality evidence supports marijuana for this condition, and the medical or recreational use of the drug is illegal in their state. Also, marijuana is still Schedule I, despite an FDA-approved derivative now on the market. How should Dr. Rogers advise his patient?

Questions for Discussion

1. What legal limits constrain Dr. Rogers in his advice to Julia?
2. If Julia plans to try marijuana anyway, how should Dr. Rogers respond? How might he protect Julia from possible side effects or complications?
3. What are the ethical considerations of off-label use of the FDA-approved product?

Case Study #12: A Forged Prescription[10]

It's been a busy Friday evening at your community pharmacy, and you are the only pharmacist on duty. Rachel S. presents at your counter with an obviously forged signature prescription for fentanyl patches for her mother. You have known Rachel and her mother Helen for years. Helen has an end-stage malignancy with metastases to bone, and doctors have told her she has less than 6 months to live. She is in excruciating pain, and the patches are the only treatment that helps.

You are very familiar with the oncologist's practice, and the signature on the prescription is clearly not his. When you confront Rachel about this, she bursts into tears. She had tried to reach the doctor's office, but it was after hours on a Friday, and there was no on-call service. In desperation, she even drove over to the office building, but it was locked, and all the cars were gone. She cannot imagine her mother having a weekend of intense, untreated cancer pain and withdrawal. How do you respond?

Questions for Discussion

1. Do you call the police?
2. What are your legal constraints?
3. What is your ethical duty? Can you help Helen in this situation?

Case Study #13: Drug Scarcity During a Crisis

During a worldwide pandemic of a viral respiratory illness, the evening news on national television announces preliminary reports that an antiviral agent typically used in the management of hepatitis may have efficacy in this new crisis. This

---

[10] This case and the three cases following it were all contributed by PharmD candidate Joshua Pearson (class of 2021).

antiviral is reported to have a significant impact on mortality and may also be useful in preventing deaths when used prophylactically. Suddenly, prescriptions for the drug are flooding your community pharmacy, and wholesalers are increasing prices as the demand skyrockets. How will you respond?

Questions for Discussion

1. Some of your regular patients with hepatitis have been taking the drug already. Others are presenting with prescriptions who are not yet sick. Who should have priority to receive the medication?
2. What ethical framework should you use to analyze this case?
3. You observe that a particular family practice physician has been writing prescriptions for all of his family members and colleagues. How should you handle this situation?
4. What are the ethical considerations of increasing prices in response to increased demand? How do patients perceive the increased costs, even when the pharmacy is not actually making any more money from the prescription?

Case Study #14: Dispensing a Recalled Medication

The manufacturer of an antiplatelet drug, used for the secondary prevention of stroke, has initiated a voluntary recall because of an undeclared carcinogenic contaminant. The company is worried about potential liability, even though there is more of the carcinogen in a single strip of bacon than in 19 years' worth of daily ingestion of the medication. The shelves of your community pharmacy were stocked with the affected lots, and you have removed them to be shipped back to the wholesaler for destruction. Ordering replacement lots will be impossible for the next 2 weeks as the wholesaler's current stock is already committed.

A patient comes in that night for a refill of his medication and is distressed to find he cannot get it. The patient is completely out. It is a holiday weekend, and he cannot reach his prescriber to order a replacement drug. You have a box full of the affected lots in the back room, and the risk of harm from the contaminant is extremely small. However, refilling a prescription with a recalled medication is illegal. Can you help your patient?

Questions for Discussion

1. How can you balance the ethical principles of beneficence with non-maleficence in this case?
2. What is your legal liability risk?
3. What is your ethical duty as a pharmacist?
4. Is there a clinically viable alternative for your patient?

Case Study #15: Who Gets the Insulin?

There is a regional insulin shortage, and you have two orders for your last insulin vial. One order is for a single 19-year-old male with type 1 diabetes. He has a history of nonadherence with his insulin regimen and has been hospitalized for diabetic ketoacidosis twice in the past 3 months. He is gainfully employed with commercial insurance. The second order is for a single 38-year-old mother with type 2 diabetes. She has five children, is unemployed, and receives public assistance. How should you respond?

Questions for Discussion
1. What ethical principles should determine who gets the last vial?
2. In considerations of priority during a medication shortage, what is the appropriate framework: social worth, medical need, ability to pay, age, or some other factor? Or should the determination simply be "first come, first served?"
3. Is there a clinically viable alternative for either patient? What barriers would need to be overcome to achieve this?

Case Study #16: When Should We Stop?[11]

Martin J. is a 38-year-old farmer who manages a 60-acre family farm with his father and two older brothers. This past Saturday was the opening day for firearm deer hunting season, and Martin was out in the woods with his 10-year-old son. Both father and son were hiding in a thicket when they saw a 12-point buck within 40 yards of their position. As the father raised his rifle to fire, he felt a sharp pain in his left chest and fell to the ground. Apparently, he had been accidentally struck by another hunter firing at the same deer.

The young boy quickly caught the attention of the other hunter, who frantically dragged Martin to his truck and drove to the nearby community hospital. Upon arrival in the emergency room, Martin was without any pulse and had only agonal respiratory attempts. The patient was quickly intubated, and a large-bore intravenous catheter was inserted. He was given large volumes of normal saline, along with Type O negative uncross-matched blood.

A chest tube was placed in the left chest, which treated the patient's large pneumothorax, but continuously drained large quantities of dark red blood. The general surgeon was called, and the patient was taken urgently to the operating room, still in shock. In the OR, the surgeon made a left thoracotomy and discovered that a high-velocity deer slug had destroyed most of the left lower lobe of his lung. A second surgeon arrived, and the two doctors performed a left lower lobectomy and closed the chest, leaving a chest tube still in place to expand the remaining left lung and drain residual fluid.

---

[11] This is partially based on an actual clinical case experienced by one of the authors (DMS).

Clinical Case Studies

Over the next four days, the patient has remained unconscious and on the ventilator. He has received 14 units of cross-matched Type A positive blood, as well as 4 units of fresh frozen plasma and 12 platelet packs. Despite these efforts, his coagulation studies remain abnormal, and he continues to bleed from the chest tube. His mean arterial blood pressure remains around 60 mm Hg (normal 70–100 mm Hg). Though both lungs are completely expanded on chest x-ray, the lung fields show evidence of acute respiratory distress syndrome. Despite 100% O_2 and high ventilator pressures, the blood pO_2 level is only 74 mm Hg (normal 100 mm Hg). Urine output is less than 500 mL. per 24 hours, indicating early renal failure.

The surgeon has approached Martin's wife, Susan. He told her that Martin is dying and that the most compassionate thing to do at this point would be to discontinue the ventilator and let him go. Susan flew into a rage and demanded that he transfer Martin somewhere else. The surgeon told her that the nearest medical center is 60 miles away and that Martin is too unstable for safe transport. Her two brothers have now joined Susan, and they are angrily threatening a lawsuit if the surgeon tries to discontinue any of the current treatments. The local church pastor is out in the waiting room, and several church members have joined him to pray for a miracle.

The ICU head nurse has asked you, the hospital ethicist, to consult, saying, "Can you calm the family down, and help us avoid a lawsuit?"

Questions for Discussion

1. What is the central ethical question in this case?
2. What is the definition of clinical futility?
3. Is it legal and ethical to discontinue the ventilator over family objections?
4. Is there any form of compromise that could help to assuage the family's anguish?

Suggested Reference George P. Smith. Medical Futility: The Template for Decision-Making. (Smith 2013).

Case Study #17: A Handshake Arrangement

Jay Smith is the 70-year-old owner-proprietor of Smith's Pharmacy in Springs Crossing,[12] a small town of about 6000 in rural Oklahoma. You are a pharmacist doing part-time work for a relief agency, but today you're filling in for Mr. Smith, who is out of town today for his own medical appointments. You're working with a technician and a clerk, both of whom have been employed at the pharmacy for over 20 years and know the operation and the patients well.

On this day, the store is relatively busy, with about 140 prescriptions filled. One patient, an older man dressed in overalls and boots, walks up to the counter and asks to talk to the pharmacist. You ask what you can do to help. The patient says, "I have a bad chest cold and need something for it." You respond, "Sure, give me a minute,

---

[12] The town of Springs Crossing, Oklahoma is fictitious.

and I'll help you find an over-the-counter product that should help." The patient looks irritated and says, "That's okay; I'll come back later when Jay is here." He then walks away. You ask the tech what that was all about, who responds, "I think he wanted something from *back here.*"

Jay Smith stops by the pharmacy near the end of the day to check on things, so you ask him about the encounter. He tells you, "Oh, yeah, I've had a long-standing agreement with the doc here in town. If I know the patient needs an antibiotic and cough syrup, then I give him an antibiotic and some cough syrup." You ask if there is a collaborative practice agreement or if the pharmacy performs any assessment (such as a rapid strep test or flu testing). Mr. Smith shrugs and says, "The doc and I work on a handshake. Patients get better, and they like getting better. I've never had a problem."

Questions for Discussion

1. What are your ethical responsibilities in this situation? What ethical principles have been violated?
2. What are your legal responsibilities?
3. Your state has "duty to report" rules for pharmacists, which include reporting any violations, attempts to violate, or aiding and abetting in the violation of pharmacy practice laws, to the board of pharmacy. How should you respond?

Suggested Reference An example of a "duty to report" rule from the State of Ohio Board of Pharmacy.[13]

Case Study #18: Discontinuing an ICD

Dorothy S. is an 81-year-old woman with a long history of cardiac disease and morbid obesity. She has survived two episodes of severe cardiac arrhythmias. The first was unobserved by any cardiac monitoring, while the second was an episode of ventricular tachycardia (VT). In both cases, she was resuscitated by paramedics and treated in the local community hospital.

After the last episode, Dorothy's cardiologist and the cardiac surgeon placed an implantable cardioverter-defibrillator (ICD) in her left chest. Such devices can cardiovert an abnormal cardiac rhythm such as VT or ventricular fibrillation (VF), preserving a patient's life (Goldstein et al. 2004; Lail and Sullivan 2017). In the 8 months since ICD placement, Dorothy has lived in reasonable comfort, despite steadily worsening heart failure and chronic shortness of breath.

In the past 2 weeks, however, Dorothy's symptoms have gotten worse. She has severe edema and shortness of breath, despite maximum therapy, including supplemental oxygen via nasal cannula and fluid restriction. Her current medications

---

[13] Pharmacist Duty to Report Requirements. State of Ohio Board of Pharmacy Web site. www.pharmacy.ohio.gov. Published 2019. Updated 10/8/2020. Accessed 12/3/2020.

include sotalol, lisinopril, furosemide, and spironolactone. Nonetheless, her last measured ejection fraction was 11%. The recent chest x-ray showed severe cardiomegaly, with the heart silhouette at 80% of the chest cavity. Dorothy's doctors have transferred her to a hospice facility, with the diagnosis of New York Heart Association Class IV congestive heart failure. A do-not-resuscitate (DNR) order is in place.

Hospice nurses are upset because the ICD keeps activating. Every time Dorothy has an episode of VT or VF, she receives an extremely unpleasant shock. Lately, such events are happening several times an hour. Nurses are concerned that the ICD is causing discomfort, and they are asking for an ethics consult. You are a member of the hospice ethics committee. Is it ethical to turn off the ICD?

Questions for Discussion

1. What additional questions would you have for the patient, the family, or the healthcare team?
2. How are ICDs different from pacemakers? Note: often, the same implantable device can combine the two functions.
3. Should ICDs be considered part of the body, just like a transplanted kidney or a replacement heart valve? Or do they substitute for a bodily function, like a ventilator or renal dialysis? Does your answer alter your ethical conclusion?
4. How does this newer technology fit into the context of hospice?
5. What ethical principles help to inform this case? Which ones should have priority?

Suggested Reference Kyle E. Karches and Daniel P. Sulmasy. Ethical Considerations for Turning off Pacemakers and Defibrillators. (Karches and Sulmasy 2015).

Policy Scenarios

Policy Case Study #1: Should the Pharmacist Support This Law?[14]

Dr. Sarah Jameson is a hospital pharmacist asked to give expert testimony before her state legislature. At issue is a proposed law that would limit the use of naloxone (Narcan®) by first responders to resuscitate patients suffering from an opioid overdose. The proposal stipulates that naloxone be used for a single individual a maximum of three times. Supporters argue that the expense of naloxone for small communities is too high; its repeated use for the same individuals interferes with the care of other patients, such as those with heart attacks, strokes, and trauma, who need urgent transport to emergency rooms. Opponents of the measure worry that such selective use of naloxone is a form of discrimination against patients with drug

---

[14] Note that this is Case #2 from Chap. 1.

addiction, a chronic disease. How should Dr. Jameson testify? Should she support or oppose this measure?

Questions for Discussion

1. What ethical principles inform this case? Think of the basic tenets of medical principlism and general ethical theories as well. The Pharmacy Code of Ethics certainly bears on this question.
2. If such a law were to pass, what might the outcomes be for patients with substance abuse disorder and the public at large (utilitarian considerations)?
3. Consider the competing duties this scenario addresses and find the duty or duties that should prevail.

Suggested References

- Beheshti A., et al. An Evaluation of Naloxone Use for Opioid Overdoses in West Virginia: A Literature Review (Beheshti et al. 2015)
- Pergolizzi JV, LeQuang JA, Taylor R, Raffa RB. The Place of Community Rescue Naloxone in a Public Health Crisis of Opioid Overdose (Pergolizzi et al. 2019)

Policy Case Study #2: Midazolam for Lethal Injection[15]

As we discussed back in Chap. 10, the possible involvement by healthcare professionals in capital punishment is ethically controversial. Lethal injection as a method was intended to mitigate concerns about Eighth Amendment prohibitions against "cruel and unusual punishment." (Stevenson and Stinneford 2020) Nonetheless, many drug manufacturers have resisted the use of their products for this purpose. For example, the U.S. company Hospira abandoned the production of sodium thiopental in Italy because of legal constraints on its use for lethal injection. The popular sedative drug is now unavailable in the U.S (Woolston 2013).

You are an academic pharmacist who also has a PhD in pharmacology. Lawyers working with the attorney general of your state have asked you to testify on their behalf to defend against a lawsuit filed by James F., a prisoner facing execution for a double homicide. The state's current lethal injection protocol uses the single drug midazolam, but lawyers for James claim it violates the Eighth Amendment because its effects are unpredictable. As an expert in pharmacology, the state feels that you can support the use of the drug and that James will never actually suffer.

Questions for Discussion

1. Can you ethically participate in this legal action to help the state execute James?
2. The state is asking for your expertise, not your direct involvement. Does that make a difference in your answer?

---

[15] Though not identical, this case is similar to Case #4 from Chap. 1, discussed in greater detail in Chap. 10.

Suggested Reference Kas K, et al. Lethal drugs in capital punishment in the USA: History, present, and future perspectives (Kas et al. 2016).

Policy Case Study #3: Should We Use Nazi Research Data?

An attendee at a scientific program of anesthesia professionals was shocked when he heard a speaker refer to the "data from Dachau" in describing the problems of treating patients with near-drowning. The speaker described "the effects of immersion in cold water upon body temperature and survival," citing Nazi experiments on condemned prisoners (Bogod 2004). There was no apparent reaction from the audience. Yet what happened next was very disturbing, when the presenter described his experiments on pigs placed in cold water:

> This slide was greeted by an audible sharp intake of breath from the audience, followed by a relieved exhalation when the speaker explained that the animal was anesthetized, "otherwise, I could never get it into the water." (Bogod 2004)

The writer of this editorial in an anesthesia journal was bothered by the lack of an emotional reaction to the abuse of death camp prisoners, compared to an obvious concern for the welfare of a pig. What possible reasons could explain the audience's response? What is your reaction?

The medical experiments performed on prisoners in the Nazi death camps during World War II have been universally condemned for their cruelty and complete abandonment of Hippocratic principles. Some examples include: forcing prisoners to drink seawater or inducing hypothermia by cold water submersion, both meant to simulate survival possibilities for downed Nazi pilots in the North Sea. Other experiments were these: the deliberate intravenous injection of live typhus organisms to compare survival times, or the induction of traumatic wounds to test sulfanilamide powders or anticoagulants.

It may be that there is some actual scientific value buried within the results obtained from these infamous and deadly experiments, but at what cost? Should we use the Nazi research data today for a reasonable medical benefit, or should it be banned?

Questions for Discussion

1. The answer to this question may well depend on your ethical point of view. Consider framing your response using a variety of ethical theories. What might Kantian ethics have to offer? Divine command theory? Utilitarianism? Virtue Ethics? How about the matter of moral complicity ("moral taint")?
2. Would you recommend a total ban on any use of the Nazi research data, or base your answer on the nature of each experiment, its value to society, or its scientific rigor?
3. These events all took place too long ago to ask for the opinion of the victims themselves. As an alternative, should we ask their descendants?

Suggested Reference Arthur Caplan: *When Medicine Went Mad: Bioethics and the Holocaust.* (Caplan 2012).

Policy Case #4: Rights of Conscience

Some pharmacists in the state of Washington feel that they have no conscience protections concerning contraceptive drugs, even those that may induce abortion. In the case of *Stormans v. Wiesman*, the owners of Ralph's Thriftway Pharmacy declined to carry certain contraceptives, such as Plan B One-Step®, that some clinicians believe may interfere with the implantation of a human embryo. Dispensing such a drug conflicts with the sincere pro-life views of Kevin Stormans and his family. In the past, when emergency contraception drugs were requested, the Stormans would simply and graciously refer the client to one of "more than 30 other pharmacies within five miles of Ralph's."[16] This type of referral arrangement is a time-honored move that had been legal and ethical in all 50 states – until now. Recently enacted Washington Board of Pharmacy rules now require dispensing of all contraceptive agents, regardless of their mechanism of action, and despite any ethical views held by pharmacists. The Stormans family sued to retain their right of conscience.

An unfavorable Ninth Circuit Court ruling led to an appeal to the U.S. Supreme Court, which on June 28, 2016, denied *certiorari*, that is, the Court refused to hear the case. In his written dissent, Justice Samuel Alito described the seeming anti-religious bias of the Court:

> This case is an ominous sign. At issue are Washington State regulations that are likely to make a pharmacist unemployable if he or she objects on religious grounds to dispensing certain prescription medications. (French 2016)

Do you share the concerns expressed by Justice Alito? Should pharmacists have a right of conscience?

Questions for Discussion

1. To claim a right of conscience, is it necessary to scientifically prove that Plan B One-Step® interferes with implantation? How does scientific evidence play a role? Note that this question does not come up in other controversial issues, such as participation in assisted suicide, where the drug in question clearly causes death.
2. When it comes to contraceptives, should states have "refusal to fill" conscience protection clauses for pharmacists?
3. What ethical principles inform the duty of a pharmacist to dispense a legal drug to a patient, despite personal misgivings?
4. If a pharmacist were to invoke a carefully-considered conscience right not to dispense a drug, how should this be handled?
5. At the time that the Supreme Court denied *certiorari* in *Stormans v. Wiesman*, the court only had eight members due to the death of Justice Antonin Scalia. If the case had gone before the court and the court had split 4–4, what would that have meant?

---

[16] Stormans, Inc., DBA Ralph's Thriftway, et al. v. John Wiesman, Secretary, Washington State Department of Health, et al. *FindLaw*. 2016. https://caselaw.findlaw.com/us-supreme-court/15-862.html#FRopinion1.6. Published 6/28/2016.

Suggested Reference Sara Rosenbaum. When Must State Public Health Laws Contain Religious Exemptions? Stormans, Inc. v. Weisman. (Rosenbaum 2016).

Policy Case #5: Non-Medical Exemptions from a Vaccine Mandate

Sometime during Christmas of 2014, a single person infected with the measles crossed paths with "tens of thousands of people at Disneyland ... including international travelers, children with still-developing immune systems, and babies too young to have been vaccinated." The result was a record outbreak of measles in the state, resulting eventually in 131 persons infected (Wick 2019).

The problem was that vaccination rates for Californians had fallen to less than the 95% required to make citizens of the state less vulnerable through herd immunity. As a result, lawmakers passed a strict mandatory vaccination law, not permitting any non-medical exemptions. The bill was effective, as the rate of vaccinations in the state climbed to 96.7% by 2017 (Wick 2019).

Nonetheless, vaccine fears and misinformation have led many children to suffer from this preventable illness (see discussion in Ch. 12). Elsewhere in the U.S., outbreaks of measles and other viral infections have arisen again, especially in states where vaccination rates remain exceptionally low. Many parents still refuse to vaccinate their children based on the mistaken impression that the targeted disease is safer than the vaccine.

As a matter of public health, what should we do to address this problem?

Questions for Discussion

1. What ethical principles are at play in this case? How might a utilitarian framework change your analysis?
2. Should parents have conscience rights concerning the immunization of their children?
3. Should the state have the power to mandate that children get vaccinated as a condition for attending public schools, even against parental wishes?
4. When does the state have a compelling interest to mandate vaccinations?
5. How should a pharmacist counsel parents who are concerned about possible adverse effects of vaccinations in their children?
6. What is the best strategy to improve public health and still preserve individual rights?
7. Suppose that the CEO of a community pharmacy chain has recently announced that they will refuse to fill prescriptions for any person not up to date on recommended vaccinations (except for those with a valid medical exemption). Is this ethical? As a pharmacist, how will you respond?

Suggested Reference Carmel Shachar and Dorit Rubinstein Reiss. When Are Vaccine Mandates Appropriate? (Shachar and Reiss 2020).

Policy Case #6: Testing a New Drug

You are the director of the pharmacy at an orthopedic specialty hospital. A new non-opioid, non-NSAID injectable analgesic has just come on the market. It is, however, relatively expensive. You are in a meeting with the pharmaceutical representative for the company selling the drug, along with Dr. John Kirk, an anesthesiologist who also runs the hospital pain management service. You're hesitant to commit to the new drug because it would add significantly to the pharmacy budget.

The pharmaceutical company representative suggests giving the new medication to all patients for one week, then using the usual analgesics the following week. The hospital could continue this weekly alternation for a few months, carefully tracking the results in terms of pain relief and adverse effects. The rep is enthusiastic that this will prove the superiority of the new analgesic. Dr. Kirk likes the idea and asks for your opinion. What do you think?

Questions for Discussion

1. Is this a workable plan, ethically and legally? What are your concerns?
2. Does this meet the definition of human research? If so, what other considerations should be accounted for?
3. Is there an alternative?
4. How should you and the anesthesiologist decide whether to adopt the new medication?

Policy Case #7: A Grant-Funded Diabetes Study

You are a new clinical faculty member for a college of pharmacy. Your colleague, Melissa Shepherd, is also a pharmacist and a Certified Diabetes Educator. She has received a grant to establish a diabetes collaborative practice clinic at a Federally Qualified Health Center (FQHC) in rural South Carolina, an impoverished area with a high prevalence of diabetes and renal disease. The grant is part of a national study of a team approach to diabetes treatment and is unique because the principal investigator is a pharmacist.

Though the patients in the area generally qualify for government health benefits such as Medicare or Medicaid, they often lack documentation, such as phone bills or tax returns, to prove that they are eligible. The FQHC offers a sliding scale payment plan based on income, but this is a difficulty for many of the patients because of their inability to prove their lack of income. Thus, many patients in the area are medically underserved even though very low cost or free care is available.

The grant is sufficient to provide for the care provided by the diabetes collaborative team, as well as any other primary care needs that participants require. To qualify, patients only need a diagnosis of diabetes or to have multiple fasting blood sugars above 110; they do not need to provide documentation of living costs or income. According to the protocol, patients receive payment for all diabetes collaborative care and primary care. However, they lose all benefits if they decline to participate or drop out of the study.

Dr. Shepherd has invited you, as a new member of the faculty, to join her team in this study, but you have a few ethical reservations.

Questions for Discussion

1. As stated, is the design of the study equitable for this patient population?
2. What, if any, are your ethical concerns?
3. Why could such a study be completely ethical in other populations, but problematic for the population in this scenario?
4. Should you join the research study team?

For Further Reading

Robert Orr: *Medical Ethics and the Faith Factor: A Handbook for Clergy and Health-Care Professionals.* Eerdmans, 2009 (Orr 2009). Dr. Orr brings a professional lifetime of clinical ethics experience to this comprehensive text filled with case studies, each containing a thorough analysis. Though written from a Christian religious perspective, secular professionals and those from other faiths will benefit from his careful and compassionate reasoning.

Robert Veatch, Amy Haddad, and EJ Last: *Case Studies in Pharmacy Ethics* (third Ed.) (Veatch et al. 2017). Oxford University Press, 2017. Dr. Veatch is the former director of the Kennedy Institute of Ethics at Georgetown University and the author or editor of over 60 books. Drs. Haddad and Last are pharmacists with both clinical and teaching experience, and Dr. Last is also a lawyer with expertise in health law. The text has an impressive number of cases, organized by topic, each one with an extensive commentary.

References

Beauchamp TL, Childress JF. Principles of biomedical ethics. 8th ed. New York: Oxford University Press; 2019.
Beheshti A, Lucas L, Dunz T, et al. An evaluation of naloxone use for opioid overdoses in West Virginia: A literature review. Am Med J. 2015;6(1):9–13.
Bogod D. The Nazi hypothermia experiments: Forbidden data? Anaesthesia. 2004;59(12):1155–6.
Caplan A. When medicine went mad: Bioethics and the holocaust. Totowa: Human Press; 2012.
Cordeiro CN, Gemignani ML. Breast Cancer in pregnancy: Avoiding fetal harm when maternal treatment is necessary. The Breast Journal. 2017;23(2):200–5.
French D. Thanks to SCOTUS, Vicious Anti-Christian State action is legal in the ninth circuit. National Review. 2016. www.nationalreview.com/corner/thanks-scotus-vicious-anti-christian-state-action-legal-ninth-circuit/. Published 6/28/2016.

Goldstein NE, Lampert R, Bradley E, Lynn J, Krumholz HM. Management of implantable cardioverter defibrillators in end-of-life care. Ann Internal Med. 2004;141(11):835–8.

Herndon CM, Nee D, Atayee RS, et al. ASHP guidelines on the Pharmacist's role in palliative and hospice care. American Journal of Health-System Pharmacy. 2016;73(17):1351–67.

Karches KE, Sulmasy DP. Ethical considerations for turning off pacemakers and defibrillators. Cardiac Electrophysiol Clin. 2015;7(3):547–55.

Kas K, Yim R, Traore S, et al. Lethal drugs in capital punishment in USA: History, present, and future perspectives. Res Soc Admin Pharm. 2016;12(6):1026–34.

Lail A, Sullivan DM. Removing implantable defibrillators at the end of life: An ethical analysis. Bioethics Network Ohio Quart. 2017;27(3)

Lamiani G, Borghi L, Argentero P. When healthcare professionals cannot do the right thing: A systematic review of moral distress and its correlates. Journal of Health Psychology. 2017;22(1):51–67.

McElrath K. Medication-assisted treatment for opioid addiction in the United States: Critique and commentary. Substance Use Misuse. 2018;53(2):334–43.

Orr RD. Medical ethics and the faith factor: A handbook for clergy and health-care professionals. Grand Rapids, Mich.: William B. Eerdmans Pub. Co.; 2009.

Pergolizzi JV, LeQuang JA, Taylor R, Raffa RB. The place of community rescue naloxone in a public health crisis of opioid overdose. Pharmacol Pharm. 2019;10(2):61–81.

Rosenbaum S. When must state public health Laws contain religious exemptions? Stormans. Inc v Weisman. Public Health Reports. 2016;131(6):847–50. https://journals.sagepub.com/doi/full/10.1177/0033354916669357

Shachar C, Reiss DR. When are vaccine mandates appropriate? AMA Journal of Ethics. 2020; https://journalofethics.ama-assn.org/article/when-are-vaccine-mandates-appropriate/2020-01. Published January, 2020

Smith GP. Medical futility: The template for decision-making. In: Palliative care and end-of-life decisions. Springer; 2013. p. 41–58.

Stevenson BA, Stinneford JF. The Eighth Amendment. Interactive Constitution. 2020, https://constitutioncenter.org/interactive-constitution/interpretation/amendment-viii/clauses/103.

Veatch RM, Haddad AM, Last EJ. Case studies in pharmacy ethics. 3rd ed. Oxford/New York: Oxford University Press; 2017.

Wick J. Essential California: How a 2014 Disneyland measles outbreak changed state history. Los Angeles Times. 2019. www.latimes.com/newsletters/la-me-ln-essential-california-20190507-story.html. Published 5/7/2019.

Woods M. Overriding parental decision to withhold treatment. AMA Journal of Ethics. 2003;5(8):247–50. https://journalofethics.ama-assn.org/article/overriding-parental-decision-withhold-treatment/2003-08

Woolston C. Death row incurs drug penalty. Nature. 2013;502(7472):417–8.

Chapter 14
Genetic Ethics and Other Cutting-Edge Issues

Having examined foundational principles, specific issues, and practical case scenarios, we now turn to the future. In this final chapter, we will consider some vexing modern questions from the arena of genetics. We will then conclude by mentioning a few newer issues that have not yet received a thorough analysis and close with a challenge to the reader.

The Genetic Frontier

April of 2003 marked a significant milestone in history: the completion of the Human Genome Project. At the cost of 2.7 billion dollars, this endeavor lasted for 13 years and was the first successful attempt to determine the human *genome*, the sequencing and mapping of all the genes of our species. Project director and geneticist Francis Collins described the accomplishment this way:

> It's a history book - a narrative of the journey of our species through time. It's a shop manual, with an incredibly detailed blueprint for building every human cell. And it's a transformative textbook of medicine, with insights that will give health care providers immense new powers to treat, prevent, and cure disease (The Human Genome Project 2020).

In a book he would later write, Dr. Collins used an even more powerful metaphor: he called the human genome "the language of God."(Collins 2007) If these comparisons are apt, then the proper oversight and application of this new technology has profound moral implications (Fig. 14.1).

Fig. 14.1 Dr. Francis Collins, announcing completion of the Human Genome Project (By National Human Genome Research Institute (NHGRI) from Bethesda, MD, USA – Francis Collins, M.D., Ph.D., CC BY 2.0, https://commons.wikimedia.org/w/index.php?curid=52360074)

Gene Editing

Almost two decades after its completion, the Human Genome Project has given us the ability to read and understand the DNA sequence within each of our 46 chromosomes, but changing it has been much more elusive. In the past, genetic therapies used viral vectors to insert modified DNA sequences into the nuclei of cells with genetic diseases. The trials were risky and dangerous and led to mixed results at best (Thomas et al. 2003; Abeel 2010; Judson 2006). But then came CRISPR, short for "Clustered Regularly Interspaced Short Palindromic Repeats." Part of a natural defense mechanism in bacteria, these are specialized stretches of DNA from previous viral attacks, a sort of immunological memory. Another bacterial defense mechanism is Cas9 (CRISPR associated protein 9), an enzyme that cuts foreign DNA (Vidyasagar 2018). In a significant "eureka" moment, researchers combined these two mechanisms into the CRISPR/Cas9 technique, a novel "gene slicing" procedure that allows for the direct editing of specific genes of any organism. Eric Lander summarizes the importance of this remarkable innovation:

> It's hard to recall a revolution that has swept biology more swiftly than CRISPR... [T]he CRISPR system – an adaptive immune system used by microbes to defend themselves against invading viruses by recording and targeting their DNA sequences – could be repurposed into a simple and reliable technique for editing, in living cells, the genomes of mammals and other organisms. CRISPR was soon adapted for a vast range of applica-

tions – creating complex animal models of human-inherited diseases and cancers; performing genome-wide screens in human cells to pinpoint the genes underlying biological processes; turning specific genes on or off; and genetically modifying plants – and is being used in thousands of labs worldwide. The prospect that CRISPR might be used to modify the human germline has stimulated international debate (Lander 2016).

Now available only in the laboratory, dreams of major clinical applications are still in the future, since research on CRISPR/Cas9 will likely take years before any human clinical treatments emerge. Yet human gene editing will likely have many significant impacts on healthcare. What ethical considerations might guide such developments? (Fig. 14.2).

Let us begin with some important practical definitions. *Somatic cell gene therapy* alters specific genes within an affected person, often in just one organ or tissue

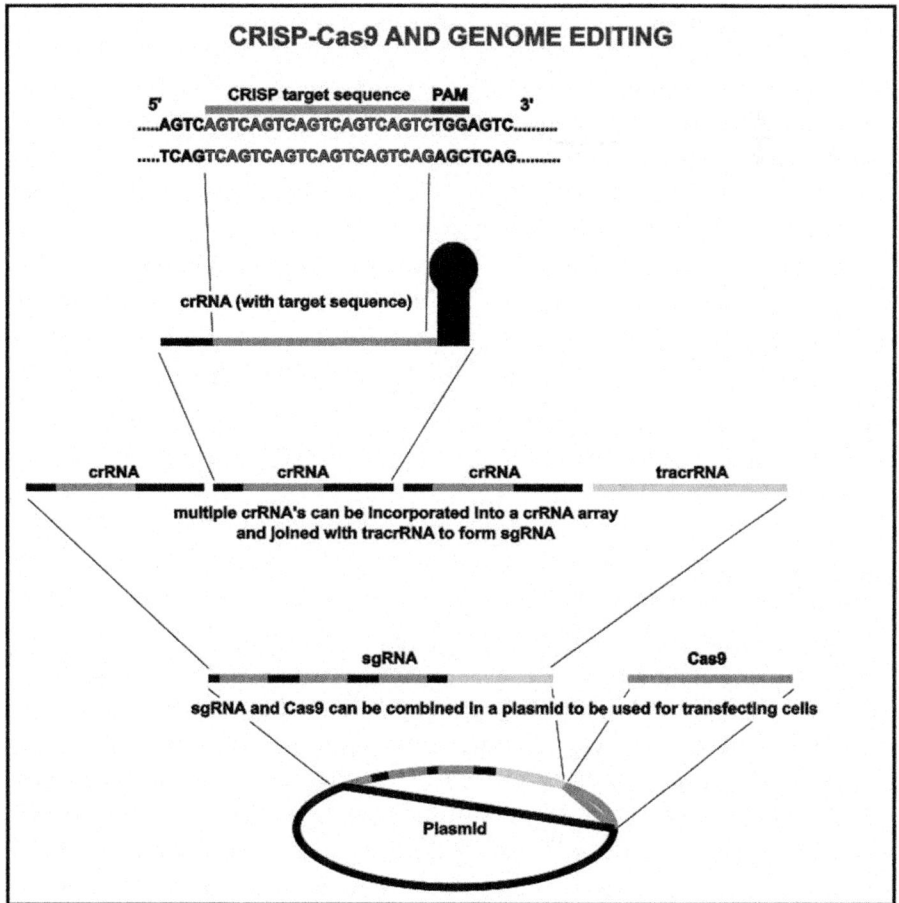

Fig. 14.2 Graphical overview of CRISPR Cas9 plasmid construction (By Kaidor – File:CRISPR overview.jpg by Nielsrca, Public Domain, https://commons.wikimedia.org/w/index.php?curid=50860715)

("somatic" comes from the Greek: *soma*, referring to the body).[1] This approach only affects the mature cells of the person's body and cannot be passed on to future generations (Lander et al. 2019). For example, viral-vector-based therapies for cystic fibrosis, affecting only the lining cells of the pulmonary bronchial system, have been tried already, and have offered modest improvement in some patients with the disease (Bosely 2015).

On the other hand, *germline treatments* focus on the entire genome of the target to permanently eliminate a particular genetic disease. Because such changes include reproductive cells, the alterations would also apply to future generations (Lander et al. 2019). There are many potential ethical concerns about using such germline therapies in human beings at this time:

- If genetically-modified human embryos are implanted and allowed to gestate, unwanted and unpredictable "downstream" side effects are likely.
- Informed consent for such experimental treatments on the affected offspring is impossible to obtain.
- Up until recently, most healthcare has centered on therapies for diseases. Germline therapies open up the possibility of "designer babies," whose genetic traits represent true human enhancement.
- Many segments of our population feel that we will be "playing God," thus meddling with human nature and human dignity.

Given the current state of research on CRISPR/Cas9, most authorities agree that germline interventions to actually give birth to genetically-altered babies should remain off-limits for now (Lander et al. 2019; Lundberg and Novak 2015; Sales 2019; Greely 2019). This consensus has persisted despite the work of He Jiankui, a researcher from Shenzhen, China, who infamously used CRISPR/Cas9 techniques to produce twin gene-edited babies, for which he received international condemnation (Li et al. 2019; Wang et al. 2019; Cyranoski and Ledford 2018).

Three-Person Embryos

Not all genetic diseases result from defects in nuclear DNA. Sometimes heritable diseases come about because of issues within mitochondrial DNA (mtDNA), the small circular portion of the human genome that resides outside of the nucleus within mitochondria (Sykes 2012). This unique property of mtDNA means that it only occurs within an oocyte, not a spermatocyte. Therefore, while both males and females can have a mitochondrial disease, only mothers can pass it on to their children (Shiel 2018). Though relatively rare, there are hundreds of mitochondrial

---

[1] Somatic. Miriam-Webster Web site. www.merriam-webster.com/dictionary/somatic. Published 2020.

The Genetic Frontier

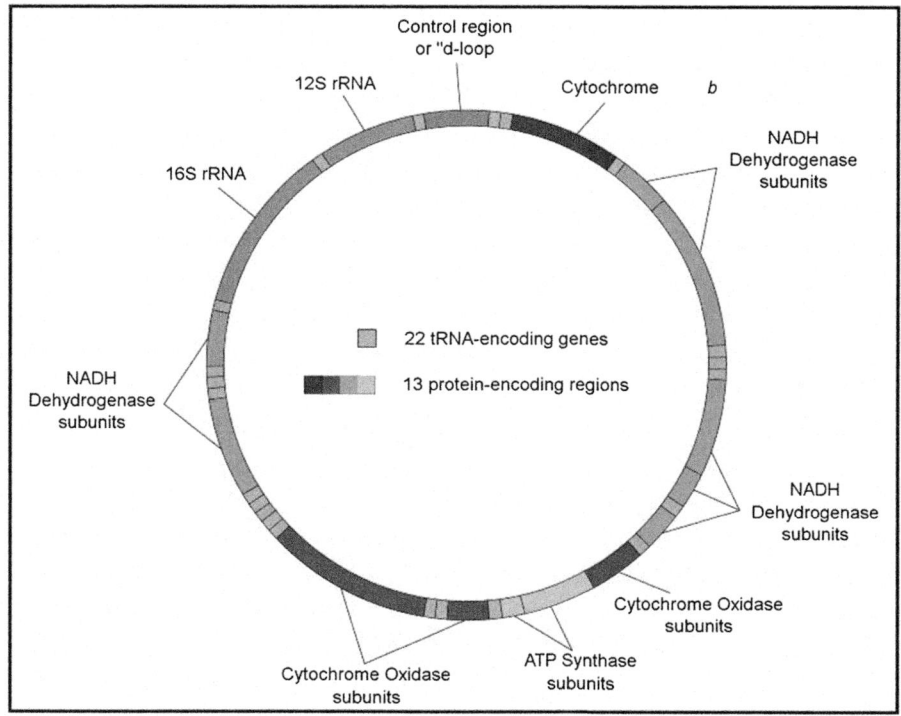

Fig. 14.3 Structure of the human mitochondrial genome (By derivative work: Shanel (talk) Mitochondrial DNA de.svg: translation by Knopfkind; layout by jhc – Mitochondrial DNA de.svg, CC BY-SA 3.0, https://commons.wikimedia.org/w/index.php?curid=4304443)

diseases, many of which have devastating consequences for those afflicted and are often fatal (Types of Mitochondrial Disease 2020) (Fig. 14.3).

Mitochondrial transfer technology is a recent breakthrough in fertility science, allowing for the creation of so-called "three-person" embryos. Using *in-vitro* fertilization (IVF) techniques, the affected mother and her male partner each contribute gametes, which laboratory clinicians then inject into an enucleated donor oocyte. A variation of this method first uses the parental oocyte and sperm to create an embryo, whose nucleus is then removed and injected into the enucleated oocyte. In either case, the enucleated donor ovum has unaffected mitochondria. If a healthy embryo results from the technique, it can be implanted in the hope of a normal pregnancy. The implanted embryo would have the complete genetic makeup of the parents, plus a small amount of donor DNA in its mitochondria. In essence, the genetic material comes from three persons (Chatterjee 2019).

Ethicists have raised several concerns with this new technology. Here are a few:

- Using donor oocytes means ovarian hyperstimulation of female volunteers, with all of the attendant medical risks, which also raises questions of coercion and exploitation of vulnerable women.

- Some forms of the technique require the destruction of one embryo for each embryo produced, which creates ethical concerns for those who hold that embryos have moral value.
- The track record of gestational surrogacy is rife with problems (see Chap. 8), and leaves "gestational mothers, egg providers, and children vulnerable to many health, legal, and social risks."(Chatterjee 2019) Mitochondrial transfer technology can only make these issues more severe.
- As with other forms of germline genetic therapies, the resultant genome in the created offspring will persist in subsequent generations, with all of the same risks discussed earlier for gene-edited germline therapies (Chatterjee 2019; Gómez-Tatay et al. 2017).

With all of these concerns, the United States and many other countries have banned the practice of mitochondrial transfer technology, though the United Kingdom has permitted it since 2015 (Haimes and Taylor 2017). You may recall our discussion in Chap. 8 of "reproductive tourism," where infertile couples travel to fertility clinics in India and elsewhere for assisted reproductive technologies, because of lower cost and fewer legal constraints. Similarly, the restrictive climate of mitochondrial transfer technology has fostered what one author calls "circumvention medical tourism." This term refers to the practice of traveling to another country to get around legal prohibitions on a medical service (Cohen 2018). Case in point: Dr. John Zhang and the world's first three-person baby.

In 2016, Dr. Zhang and colleagues from the New Hope Fertility Center in New York City used mitochondrial transfer to help a Jordanian couple have a child free of Leigh syndrome, a severe neurological disease caused by a mitochondrial defect. The parents had already suffered four miscarriages and had lost two young children from the illness. The American medical team traveled to Mexico to perform the procedure, where the legal climate is more agreeable (Roberts 2016). Though the newborn baby boy appears healthy, such circumvention of legal constraints against this type of procedure only heightens the aforementioned ethical concerns. Mitochondrial transfer technology remains highly controversial and is still illegal in the United States.

Genetic Information: Testing and Privacy Issues

The modern explosion of genetic information has created significant shifts in clinical practices and has raised some serious ethical questions as well. As a simple example, consider the breast cancer susceptibility genes (BRCA) 1 and 2, mutations of which have been known since the 1990s to increase the risk of gynecologic malignancies. Roughly 1 in 500 women have a mutation in BRCA1 or BRCA2, conferring a 50% lifetime risk of breast cancer by age 70, and a 30% risk of ovarian cancer (Sarin 2006; BRCA 2019). In these high-risk patients, many elect chemoprevention with tamoxifen or other estrogen modulators, and may even undergo prophylactic surgery, including bilateral mastectomies and oophorectomies (Liede et al. 2017).

Specific to BRCA testing, but emblematic of genetic testing in general, a variety of ethical concerns have arisen. Many at-risk women are afraid to be tested for fear of discrimination by insurance companies or employers. For this reason, the Genetic Information Non-Discrimination Act (GINA), passed in 2008, bans insurance companies from denying coverage or raising premiums based on personal or family genetic information (Hampton 2008).

Today, whole-genome and exome sequencing are routine practices that provide detailed genetic information and may uncover a genetic cause or predisposition for many pathological conditions. From the National Institutes of Health:

> With next-generation sequencing, it is now feasible to sequence large amounts of DNA, for instance all the pieces of an individual's DNA that provide instructions for making proteins. These pieces, called exons, are thought to make up 1 percent of a person's genome. Together, all the exons in a genome are known as the exome, and the method of sequencing them is known as whole-exome sequencing. This method allows variations in the protein-coding region of any gene to be identified, rather than in only a select few genes. Because most known mutations that cause disease occur in exons, whole exome sequencing is thought to be an efficient method to identify possible disease-causing mutations (U.S. National Library of Medicine 2020).

Given all of this, a bewildering array of genetic testing options is currently available. Choices range from simple susceptibility testing, such as for BRCA, to prenatal genetic testing of expectant mothers to guide decisions about pregnancy termination, to whole-genome and exome sequencing in pediatrics to discover if a presenting clinical syndrome has a genetic basis. These new genetic horizons present several ethical concerns:

- Clinical competency: a recent survey of primary care providers demonstrated that "they felt unprepared to work with patients at high risk for genetic conditions and were not confident about interpreting test results."(Hauser et al. 2018)
- Informed consent: clinicians should educate their patients on reasonable expectations. Patients who consent to undergo testing may be unprepared for the wealth of information produced, some of which may be unexpected, embarrassing, or upsetting (Braverman et al. 2018).
- Risks to minors: predictive testing of minors requires parental consent. Potential adverse risks include: "psychological harm to the minor being tested, negative effects on the family as a whole, risk of social discrimination and restriction, as well as violation of future autonomy." Recent surveys have been more positive, claiming that decreasing uncertainty about future risks confers psychological benefit and helps families to plan for the future (Braverman et al. 2018) (Fig. 14.4).

In the arena of genetic testing, direct to consumer (DTC) testing presents an additional ethical and regulatory challenge. At the beginning (in 2006), these tests initially dealt with nutrition, and the companies often made various claims not yet verified by research (Seward 2018). According to the Federal Trade Commission, "the results of their DTC genetic tests often include dietary advice and sales offers

Fig. 14.4 Logo of 23andMe, a personal genomics and biotechnology company. (By 23andMe – https://auth.23andme.com/app/auth/static/img/ttam_name_logo.d1dbc0855e13.svg, Public Domain, https://commons.wikimedia.org/w/index.php?curid=28805093)

for 'customized' dietary supplements and cosmetics." (Direct-to-Consumer Genetic Tests 2020) The following years have led to many more robust and specific claims, including the prediction of risks for Alzheimer's disease, arthritis, asthma, genes that influence athletic performance, various cancers, cardiovascular disease, celiac disease, macular degeneration, multiple sclerosis, and many others (Caulfield et al. 2010). There are many ethical risks to these kinds of tests and the claims based on them:

- Though often based on research findings, most DTC testing has not undergone rigorous clinical evaluation.
- The tests are directly available to the public via the Internet, without the involvement of a healthcare professional and not as part of a comprehensive clinical workup.
- The actual value of such tests remains unclear, and may often be exorbitantly expensive (Seward 2018; Direct-to-Consumer Genetic Tests 2020; Caulfield et al. 2010).

Furthermore, is the genetic data collected by DTC genetic testing part of a patient's private health information and protected by HIPAA? Even though HIPAA privacy rules were strengthened in 2013 to include genetic data, this may still be inadequate. The reason is simply that many DTC companies are not healthcare providers (to whom HIPAA would apply), and those who purchase their products are customers or consumers, not their patients. Therefore, DTC corporations may not feel obligated to store their data securely, and most compile the information into large databases for sale to research companies (Seward 2018). Such practices only serve to heighten the privacy and security concerns.

Precision Medicine

Precision medicine, sometimes called "personalized medicine," is "the ability to use molecular data to target therapies to patients for whom they offer the most benefit at the least risk."(Olson 2017) The driving force is *pharmacogenomics*, the study of how inherited genetic differences may affect an individual patient's response to specific medications (What is pharmacogenomics? 2020). While precision medicine encompasses more than pharmacogenomics, we'll briefly consider the case of warfarin as an example of how the clinical problems posed by precision medicine can also lead to ethical concerns.

Genetic variants of the primary enzyme responsible for the metabolism of S-warfarin, *CYP2C9*, cause pharmacokinetic changes in the more potent of the two enantiomers, prolonging the half-life and increasing both the effect and the risk of bleeding (Sanderson et al. 2005). Also, genetic variants in *VKORC1*, the rate-limiting enzyme in the oxidation-reduction cycling of vitamin K-dependent clotting factors, can increase the pharmacodynamic response to warfarin, thus reducing the dosage required for a therapeutic INR (Perlstein et al. 2012). Patients with both *CYP2C9* and *VKORC1* variants can be very sensitive to warfarin. They may, therefore, require very low doses of the drug for effective treatment. The FDA has required that information on the dosing requirements and bleeding risk associated with *CYP2C9* and *VKORC1* variants be added to warfarin package information (Roth 2008). In short, warfarin dosing based on pharmacogenomics information appears to offer an advancement in therapy and may reduce the need for tight monitoring (Gulseth et al. 2009).

However, there is a practical problem with this approach. Genotyping is still not widely available clinically, and samples are often sent to reference laboratories, with delays in the results by as much as a week. In the meantime, the vast majority of patients will have had their warfarin dosage adjusted based on clinical judgment. It is therefore not clear that pharmacogenomic warfarin dosing is superior to clinical judgment (Grossniklaus 2010; Drozda et al. 2018).

The essential ethical question raised by all of this is that of *clinical equipoise*. You may recall that we defined clinical equipoise in research as a state of uncertainty (Chap. 6), where the researchers do not know if a given treatment is helpful or not. Despite a genotypic association with dosing requirements and bleeding risk, pharmacogenomic dosing has not yet proven superior to dosing by a well-trained clinician. This issue is the burden for proponents of precision medicine: it is still an open research question, at least where clinical equipoise still exists. Any attempt to claim otherwise is premature.

The same problem as seen in warfarin dosing is also evident in cancer chemotherapy protocols, traditionally built on dosing agents by body surface area. Would dosing based on genotype be superior? This question is still open for further study.

Addressing these complex clinical problems will require extensive research, employing vast amounts of data collection to gather information on individual genotypes and matching them to phenotypes (i.e., to the pharmacokinetic and pharmacodynamic responses). The ultimate question, of course, will be whether precision medicine actually improves outcomes.

Along the way, additional ethical questions arise: Who will manage this data? How will privacy be assured? Is there a way to make the data "open-source" for research, or will each database be isolated from all the others? Precision medicine is still in its infancy and raises a significant number of ethical concerns, as well as many potential issues we cannot yet foresee.

Additional Challenges in Genetic Ethics

As we conclude this section on genetic ethics, here are a few additional areas of concern:

- *Mandatory genetic screening of newborns.* How will we balance parental autonomy against unrealistic parental expectations?
- *Genetic testing for disease risk.* The tests are expensive, and this creates issues of discrimination. Will such testing only be available to higher socio-economic classes?
- *The use of archived biospecimens.* Researchers are increasingly using blood and tissue samples for genetic studies, often for purposes that go beyond that for which patients initially gave consent.[2] Should this information be in the public domain, or are there privacy concerns?

Other Cutting-Edge Issues

In this final section, we will introduce a few remaining areas of ethical controversy that have not yet received much detailed analysis.

Machine Learning and Artificial Intelligence (AI)

What is the role of AI in clinical decision-making? One issue (among many) is the "black box" problem:

> [I]t isn't always possible to know how an AI system reached its results – the so-called black box problem. Advanced machine-learning technologies, which can absorb large quantities of data and identify statistical patterns often without explicit instruction, can be especially hard to verify. Physicians who blindly follow such an inscrutable system might unintentionally harm patients (Rethinking Medical Ethics 2019).

Furthermore, computer algorithms may reflect certain unconscious racial or gender biases, which might persist as outright discrimination. Who is liable for AI-induced errors? How will patient privacy be protected? (Rethinking Medical Ethics 2019) For pharmacists, how will counseling be handled if medication therapy is machine-based? How much of the process can be safely automated? (Fig. 14.5)

---

[2] Contemporary Issues for Protecting Patients in Cancer Research: Workshop Summary. Paper presented at: National Cancer Policy Forum 2014; Washington, DC.

Fig. 14.5 A black box (in the circuit analysis sense), showing input and output (By Frap at English Wikipedia – Transferred from en.wikipedia to Commons, Public Domain, https://commons.wikimedia.org/w/index.php?curid=4605365)

The Rise of Telehealth and Telemedicine

The Centers for Medicare & Medicaid Services define telemedicine as "two-way, real-time interactive communication between the patient and the physician or practitioner at the distant site." (Telemedicine 2020) Pharmacists will undoubtedly be involved in these strategies, designed for patient convenience and reduced cost. However, these methods raise significant questions of privacy, confidentiality, and continuity of care (Chaet et al. 2017). What are the ethical implications of "telepharmacy" and "tele-medication therapy management?"

The Ethics of Pharmacy Benefit Managers (PBMs)

PBMs "provide added value and reduced costs by specifically managing the pharmaceutical benefit on behalf of a plan sponsor" (a health insurance company). Some questionably ethical practices include giving more expensive drugs preferred status, costing patients more, and so-called "gag" clauses in contracts, which prohibit pharmacists from informing their patients of less-expensive ways of accessing their prescription medications (Drettwan and Kjos 2019). These concerns have become more strident as some medications on the market for a long time have increased in price, since they were only available from a single manufacturer, and as many PBMs moved certain generics to the second- and third-tier, which has increased co-pays. Considerable anti-PBM sentiment is growing in society generally. For example:

> States and Congress have taken swift action on gag clauses. Between 2016 and September 2018, 27 states enacted laws that sought to prevent gag clauses. In September 2018, Congress passed a federal law prohibiting gag clauses (Seeley and Kesselheim 2019).

In spite of these changes, ethical controversy remains. What role should pharmacists and pharmacy organizations play in shaping the future of prescription payment mechanisms?

Mandatory Preventative Treatments and Health Behaviors

As discussed in Chap. 12, pandemics and other regional health crises may place enormous pressures on governing bodies (local, state, and federal) to pass laws mandating the use of vaccines, novel therapies, the wearing of face masks, and various shelter-in-place regulations. All of these dictates, though intended for the greater public good, may infringe on individual freedoms and personal autonomy.

What duties do pharmacists have to educate the public on public health measures to enhance adherence? In the recent COVID-19 pandemic, many chain pharmacies were slow to adopt social distancing and the use of personal protective equipment. Clinicians must study the reasons for this and explore the ethical duties of pharmacies for both their personnel and their patients. Furthermore, given the visibility of the profession, what moral obligation do pharmacists have to act as role models for reasonable public health measures? As you explore these questions, be sure to remember that public health recommendations may change during a challenging public crisis, as new information arises and the situation evolves. Ethically-aware clinicians will be debating how to balance personal autonomy versus public health for a long time to come.

The Continued Opioid Crisis

This is not exactly "cutting-edge," but the epidemic of addiction and overdoses is an ongoing crisis created by the unethical practices of drug manufacturers and prescribers, at times with the tacit collusion of pharmacists, that will be with us well into the future. It is now incumbent on pharmacy professionals to fulfill their fiduciary duty to patients and help make things right. This crisis creates a tough conflict of competing duties for those of us on the front lines: providing compassionate care for patients who need treatment of chronic pain, while at the same time preventing abuse and diversion, the most common sources of opioid-related problems in pharmacies (Stratton et al. 2018). Community pharmacists must be skilled in tracking morphine-equivalent doses of opioids, correctly dealing with forged prescriptions, using state-based controlled drug monitoring programs, and implementing proper disposal of controlled substances (Pharmacists' Role 2014).

Conclusion

At the beginning of this text, we pointed out that pharmacy ethics is a relatively new discipline, but because of an increased clinical role for pharmacists, it has growing importance. In teaching the foundational principles of this discipline, we have endeavored to place high trust in the moral judgment of pharmacy students and practitioners. Our goal has always been to teach the reader *how* to think, based on ethical principles and moral reasoning, not necessarily *what* to think.

Clear moral thinking utilizes the framework of normative ethics and competing duties, recognizing that a clinician cannot fulfill all apparent duties within a given situation. Deciding which duties should prevail is the balancing act of normative ethics. This text has presented a clear-cut pathway for resolving ethical dilemmas based on foundational principles and critical thinking.

Though we have deeply considered many types of clinical controversies and dilemmas, this chapter has shown that many more ethical problems remain. We now turn to you, the reader, to master the principles and examples contained herein, so that you may apply them in clinical practice. It is our sincere hope that you will gain wisdom and clinical expertise to bring about physical, mental, and emotional healing and well-being for your patients.

References

Abeel S. Smoke and mirrors: Jesse Gelsinger, human experimentation, and gene therapy. Int J Human. 2010;8(4):15–30.

Association AP. Pharmacists' role in addressing opioid abuse, addiction, and diversion. J Am Pharm Assoc. 2014;54(1):e5–e15.

Bosely S. Gene therapy treatment for cystic fibrosis may be possible by 2020, scientists say. The Guardian. 2015. http://www.theguardian.com/science/2015/jul/03/gene-therapy-cystic-fibrosis-2020-scientists. Published July 2015.

Braverman G, Shapiro ZE, Bernstein JA. Ethical issues in contemporary clinical genetics. Mayo Clinic Proc Innov Qual Outcome. 2018;2(2):81–90.

BRCA Gene mutations. Center for Disease Control and Prevention. 2019. www.cdc.gov/cancer/breast/young_women/bringyourbrave/hereditary_breast_cancer/brca_gene_mutations.htm. Published 5 April 2019.

Caulfield T, Ries N, Ray P, Shuman C, Wilson B. Direct-to-consumer genetic testing: good, bad or benign? Clin Genet. 2010;77(2):101–5.

Chaet D, Clearfield R, Sabin JE, Skimming K. Ethical practice in telehealth and telemedicine. J Gen Intern Med. 2017;32(10):1136–40.

Chatterjee A. Three-person IVF: from genetic disease to genetic design? Center for Genetics and Society. www.geneticsandsociety.org/biopolitical-times/three-person-ivf-genetic-disease-genetic-design. Published 26 March 2019.

Cohen G. Circumvention medical tourism and cutting edge medicine: the case of mitochondrial replacement therapy. Ind J Glob Leg Stud. 2018;25(1):439–62.

Collins FS. The language of God: a scientist presents evidence for belief. Waterville: Wheeler Publishers; 2007.

Cyranoski D, Ledford H. Genome-edited baby claim provokes international outcry. Nature. 2018;563(7733):607–8.

Direct-to-Consumer Genetic Tests. Federal Trade Commission Consumer Information Web site. www.consumer.ftc.gov/articles/0166-direct-consumer-genetic-tests. Published 2018. Accessed 27 Nov 2020.

Drettwan JJ, Kjos AL. An ethical analysis of Pharmacy Benefit Manager (PBM) practices. Pharmacy. 2019;7(2):65.

Drozda K, Pacanowski MA, Grimstein C, Zineh I. Pharmacogenetic labeling of FDA-approved drugs: a regulatory retrospective. JACC: Bas Trans Sci. 2018;3(4):545–9.

Gómez-Tatay L, Hernández-Andreu JM, Aznar J. Mitochondrial modification techniques and ethical issues. J Clin Med. 2017;6(3):25.

Greely HT. How should science respond to CRISPR'd babies? Issues Sci Technol. 2019;35(3):32–7.

Grossniklaus D. Testing of VKORC1 and CYP2C9 alleles to guide warfarin dosing: test category: pharmacogenomic (treatment). PLoS Curr. 2010;2

Gulseth MP, Grice GR, Dager WE. Pharmacogenomics of warfarin: uncovering a piece of the warfarin mystery. Am J Health Syst Pharm. 2009;66(2):123–33.

Haimes E, Taylor K. Sharpening the cutting edge: additional considerations for the UK debates on embryonic interventions for mitochondrial diseases. Life Sci Soc Policy. 2017;13(1):1–25.

Hampton T. Congress passes bill to ban discrimination based on individuals' genetic makeup. JAMA. 2008;299(21):2493.

Hauser D, Obeng AO, Fei K, Ramos MA, Horowitz CR. Views of primary care providers on testing patients for genetic risks for common chronic diseases. Health Aff. 2018;37(5):793–800.

Judson HF. The glimmering promise of gene therapy. Technol Rev Manchester. 2006;109(5):40.

Lander ES. The heroes of CRISPR. Cell. 2016;164(1-2):18–28.

Lander ES, Baylis F, Zhang F, et al. Adopt a moratorium on heritable genome editing. Nature. 2019;567(7747):165–8. https://doi.org/10.1038/d41586-019-00726-5.

Li J-r, Walker S, J-b N, X-q Z. Experiments that led to the first gene-edited babies: the ethical failings and the urgent need for better governance. J Zhejiang Univ Sci B. 2019;20(1):32–8.

Liede A, Mansfield CA, Metcalfe KA, et al. Preferences for breast cancer risk reduction among BRCA1/BRCA2 mutation carriers: a discrete-choice experiment. Breast Cancer Res Treat. 2017;165(2):433–44.

Lundberg AS, Novak R. CRISPR-Cas gene editing to cure serious diseases: treat the patient, not the germ line. Am J Bioeth. 2015;15(12):38–40.

Olson MV. Precision medicine at the crossroads. Hum Genomics. 2017;11(1):23.

Perlstein TS, Goldhaber SZ, Nelson K, et al. The creating an optimal warfarin nomogram (CROWN) study. Thromb Haemost. 2012;107(01):59–68.

Rethinking Medical Ethics. Forbes Magazine. www.forbes.com/sites/insights-intelai/2019/02/11/rethinking-medical-ethics/#22480f086f03. Published 11 February 2019

Roberts M. First 'three person baby' born using new method. BBC News Online. www.bbc.com/news/health-37485263?utm_medium=email&utm_source=digg. Published 27 September 2016.

Roth M. The Warfarin revised package insert: is the information in the label too thin. Houst J Heal Law Policy. 2008;9:279.

Sales L. Leading gene editing experts call for a moratorium on its use in humans. Chain React. 2019;135:22.

Sanderson S, Emery J, Higgins J. CYP2C9 gene variants, drug dose, and bleeding risk in warfarin-treated patients: a HuGEnet™ systematic review and meta-analysis. Genet Med. 2005;7(2):97–104.

Sarin R. A decade of discovery of BRCA1 and BRCA2: are we turning the tide against hereditary breast cancers? J Cancer Res Ther. 2006;2(4):157.

Seeley E, Kesselheim AS. Pharmacy benefit managers: practices, controversies, and what lies ahead. Issue Brief (Commonw Fund). 2019;2019:1–11.

Seward B. Direct-to-consumer genetic testing. Therap Innov Regul Sci. 2018;52(4):482–8.

Shiel W. Medical definition of mitochondrial inheritance. MedicineNet. www.medicinenet.com/script/main/art.asp?articlekey=4402. Published 12 December 2018.

Stratton TP, Palombi L, Blue H, Schneiderhan ME. Ethical dimensions of the prescription opioid abuse crisis. Bull Am Soc Hosp Pharm. 2018;75(15):1145–50.

Sykes B. Mitochondrial DNA and human history. The Human Genome Wellcome Trust. 2012. https://web.archive.org/web/20150907140051/http://genome.wellcome.ac.uk/doc_WTD020876.html

Telemedicine. Medicaid.gov Website. www.medicaid.gov/medicaid/benefits/telemedicine/index.html. Published 2020. Accessed 29 Dec 2020

The Human Genome Project. National Institutes of Health Web site. www.genome.gov/human-genome-project. Published 2019. Accessed 22 Dec 2020.

Thomas CE, Ehrhardt A, Kay MA. Progress and problems with the use of viral vectors for gene therapy. Nat Rev Genet. 2003;4(5):346–58.

Types of Mitochondrial Disease. United Mitochondrial Diseae Foundation Web site. www.umdf. org/types/. Accessed 25 Nov 2020.

What are whole exome sequencing and whole genome sequencing? U.S. National Library of Medicine Web site. https://ghr.nlm.nih.gov/primer/testing/sequencing. Published 2020. Accessed 26 Nov 2020.

Vidyasagar A. What is CRISPR? LiveScience. 2018. www.livescience.com/58790-crispr-explained.html. Published 21 April 2018.

Wang C, Zhai X, Zhang X, Li L, Wang J. Liu D-p. gene-edited babies: Chinese Academy of Medical Sciences' response and action. Lancet. 2019;393(10166):25–6.

What is pharmacogenomics? U.S. National Library of Medicine Web site. https://ghr.nlm.nih.gov/primer/genomicresearch/pharmacogenomics. Published 2020. Accessed 9 Dec 2020.

Glossary

Abortifacient a medication that causes the loss of embryonic or fetal life *in utero*.

Advance directive a written document expressing a patient's wishes at the end of life, e.g., a living will or durable power of attorney.

After-birth abortion a controversial synonym for infanticide.

Allocation criteria a set of guidelines for the distribution of scarce resources during a pandemic of other healthcare crisis.

Altruistic surrogacy an informal arrangement wherein a woman, often a family member, agrees to gestate a child on behalf of another couple without payment other than medical expenses (cp. **commercial surrogacy**).

Apothecary an outdated term for a person who prepares and sells medications to the public, now supplanted by the word 'pharmacist.'

Arbeit macht frei a German expression that translates, "work makes you free," which appeared in signs over the entrances to Auschwitz and other Nazi death camps.

Artificial insemination a fertility treatment wherein sperm collected from the male partner or a donor are instilled into the female vagina or uterus (cp. **intrauterine insemination**).

Artificial intelligence (AI) a general term for computer-assisted technologies designed to perform tasks that usually require human input.

Assent (pediatrics) agreement to a treatment or procedure by a minor who lacks decision-making capacity.

Assisted reproductive technologies (ART) a general term for *in vitro* fertilization and other techniques to help infertile couples to conceive and bear children.

Assisted suicide (physician-assisted suicide, PAS) the taking of one's own life, usually in the context of a terminal illness, with the aid of a physician.

Autonomy the right to make healthcare decisions for oneself, free of coercion or constraint.

Banality of evil a phrase coined by political philosopher Hanna Arendt to describe the bureaucratization of Nazi abuses by Adolf Eichmann and others.

Belmont Report landmark 1979 set of guidelines that established three fundamental principles for the ethical governance of research on human subjects: respect for persons, beneficence, and justice.
Beneficence in healthcare, the ethical obligation to act in the best interests of one's patients; one of the four components of medical principlism.
Best interests standard of consent (implied consent) when an incapacitated patient's preferences are unknown and no surrogate is available, a decision to do what most patients would choose under similar circumstances.
Biological mother the genetically-related mother of a child.
Black box problem the unintended patient harm that might result from AI or machine-learning systems where the basis for a clinical decision is unknown.
Brain death (death by neurological criteria) also referred to as the "Harvard criteria," a definition of death established by the irreversible cessation of all functions of the entire brain, including the brain stem.
Breakthrough ovulation ovulation that occurs despite using hormonal contraception.
Breast cancer susceptibility genes (e.g., BRCA 1 and 2) a set of genetic markers that carry a significantly increased lifetime risk in women for breast and ovarian malignancies.
Cardiopulmonary death a definition of death based on the cessation of cardiac and lung activity.
Categorical imperative an ethical rule proposed by Immanuel Kant. Version 1 states one should only perform an action only if we could make it a binding rule for everyone. Version 2 states that persons should be ends in and of themselves, never the means to another person's ends (see **hypothetical imperative**).
Circumvention medical tourism the practice of traveling to another country to get around legal prohibitions for a medical service (cp. **reproductive tourism**).
Clinical equipoise the idea that genuine therapeutic uncertainty must be present for a randomized, placebo-controlled clinical trial to be ethical.
Collaborative practice agreement (CPA) a practice arrangement that allows physicians and pharmacists to share responsibilities and treatment authority for many disease entities.
Combined oral contraceptive an orally-administered contraceptive that uses a combination of a progestin and an estrogen to inhibit follicular development and ovulation.
Commercial surrogacy **a formal (contractual) arrangement wherein a woman agrees to gestate a child on behalf of another couple. Such an arrangement typically involves payment in addition to covering medical expenses (cp. altruistic surrogacy).**
Commodification of human beings a critique of some forms of assisted reproductive technologies, wherein the resulting child is seen more as a product than a person.
Competency having the legal authority to direct one's own financial and personal affairs, as established by a court of law (cp. **decision-making capacity**).

Competing duties a set of apparent clinical duties whose relative weight must be decided on in ethical decision-making.

Complementary, alternative, or integrative medicine (CAM) therapies non-mainstream health-promotion practices used in addition to, in place of, or coordinated with standard treatments.

Confidentiality in healthcare, the principle of keeping patient information private.

Conflation fallacy commonly referred to as "comparing apples to oranges," to draw false conclusions from an inaccurate or inappropriate comparison.

Consequentialist reasoning ethical reasoning based on results or outcomes, e.g., classical utilitarianism.

Contraceptive effect in discussing the action of a contraceptive, refers to any action that helps to prevent follicular development or subsequent release of a secondary oocyte.

Corresponding responsibility the legal standard whereby a pharmacist shares equally with the prescriber the culpability for filling inappropriate or illegal prescriptions of controlled substances.

Covenant a binding promise or pledge that is stronger than a contract and implies a fiduciary relationship.

CRISPR (clustered regularly interspaced short palindromic repeats)/Cas9 (CRISPR-associated protein 9) an advanced method for editing DNA of biological organisms.

Cryopreservation of embryos the "freezing" of human embryos to preserve them for later use for reproduction or experimentation.

Dead donor rule the ethical principle that the harvesting of a donor organ for transplantation cannot cause the death of the organ donor.

Death by neurological criteria (see **brain death**)

Decision-making capacity the ability of an adult patient to make informed treatment decisions, as established by a clinician (cp. **competency**).

Deontological reasoning ethical reasoning based on a set of principles of guidelines, independent of results.

Dilation and evacuation a technique for a second-trimester surgical abortion, *not* the same as the now illegal method called intact dilation and extraction ("partial-birth abortion").

Direct to consumer (DTC) genetic testing genetic analyses made directly available to the public without the involvement of a healthcare professional.

Distributive justice the principle of treating all patients equally, regardless of age, biological sex, gender, sexual orientation, ethnicity, race, citizenship, religion, social class, or other medically non-relevant factors; one of the four components of medical principlism.

Divine command theory any of a range of ethical theories based on an authoritative religious text, e.g., the Torah, the Koran, or the Christian Bible.

Double effect (principle of) an ethical concept that emphasizes *intent* in the face of a treatment or medication that may have both beneficial and harmful effects. If the intent of the intervention is medically reasonable and beneficent, then the use of it is justified.

Durable power of attorney for health care (DPAHC) also called a healthcare proxy. This document designates another person, often a family member, to make healthcare decisions on a person's behalf if they become incapacitated.

Ectopic pregnancy a pregnancy where the embryo has implanted outside of the uterus, typically in one of the Fallopian tubes. Rupture of an ectopic pregnancy is a life-threatening surgical emergency.

Eighth Amendment part of the Bill of Rights, forbids the inflicting of "cruel and unusual punishments."

Emancipated minor a minor that lives independently of parents, due to marriage or service in the military (a legally-determined category). Such an individual is responsible for his or her own care and may be treated without parental consent.

Embryo donation also called "embryo adoption," the giving of a couple's cryopreserved embryo to another infertile family so that they may conceive.

Emergency contraception the use of a one-time contraceptive agent to suppress or delay ovulation after unprotected intercourse (e.g., see **levonorgestrel**).

Empirical functionalism the view that a set of functions or abilities defines human personhood.

Enlightenment a seventeenth and eighteenth century movement in Europe that focused on reason and rejected tradition, especially that of religious authorities.

Epidemic a widespread breakout of an infectious disease.

Ethics the study of principles of right and wrong behavior.

Eugenics (positive and negative) from Greek roots for "good birth," eugenics uses ideas of genetics and hereditary to improve human society. Positive eugenics encourages the development of desirable traits, while negative eugenics discourages the development of undesirable traits.

Euthanasia also called "mercy killing," the direct taking of a human life by a clinician to relieve suffering.

Feeblemindedness a vague "diagnosis" from the eugenics era, roughly describing someone with a cognitive disability.

Fiduciary relationship a legal or ethical relationship of deep trust and confidence with another person.

Fourteenth Amendment one of three constitutional amendments passed at the end of the U.S. Civil War, granting full citizenship to African-Americans. In particular, it guarantees that no person shall be deprived of life, liberty, or property without due process of law.

Fraudulent treatments treatments known to be ineffective or harmful (see **medical quackery**).

Freedom of action principle free moral agents may choose what to do in a given situation. An inability to act (e.g., due to a physical limitation) is not a moral failure.

Gene editing the alteration of DNA, typically employing CRISPR/Cas9 technology.

Genetic Information Non-Discrimination Act (GINA) a federal law passed in 2008 that bans insurance companies from denying coverage or raising premiums based on personal or family genetic information.

Germline gene therapy modification of the DNA of whole embryos or reproductive cells, with changes to the genome that can be passed on to future generations.

Gestational mother the woman whose uterus gestates a fetus, whether the fetus is from natural reproduction or reproductive technology.

Golden mean according to Aristotle, the theoretical basis of virtue ethics, where a given virtue is defined as the middle ground between two extremes.

Gonzales v. Carhart a landmark 2007 decision by the U.S. Supreme Court that upheld as constitutional the Partial-Birth Abortion Ban Act of 2003.

Griswold v. Connecticut a landmark 1965 decision by the U.S. Supreme Court that established the right of couples to buy and use contraceptives without any government restriction.

H1N1 sometimes referred to as the "swine flu," a subtype of influenza virus that caused a major pandemic recently in 2009.

Healthcare proxy a surrogate decision-maker for a patient lacking decision-making capacity (see **durable power of attorney for healthcare**).

Herd immunity a term used primarily in relation to vaccination practices. When most members of a community are immune to an infectious disease, non-vaccinated or non-immune members are less likely to contract the disease.

HIPAA the Health Insurance Portability and Accountability Act of 1996. Among other provisions, it provides legal protection for the confidentiality of patient medical information.

Hospice care care directed toward the relief of symptoms and improving quality of life in a patient with a terminal illness and typically with less than six months to live.

Human dignity an affirmation of individual human worth. Depending on its competing definitions, dignity may be intrinsic to all human beings or it may be diminished or lost through illness and suffering.

Human Genome Project the 13-year research project, completed in 2003, that successfully sequenced and mapped all of the genes of the human species.

Humoral theory of disease an outmoded theory, held by Hippocrates and other ancient Greek physicians, that taught that diseases arose by an imbalance of four humors (blood, yellow bile, black bile, and phlegm).

Hypothetical imperative a conditional if/then statement of how to accomplish a personal goal, which Immanuel Kant would call mere prudence, rather than an ethical rule (see **categorical imperative**).

Imminent death in a terminally-ill patient, when death is expected soon, usually in days or a few weeks.

Implantable cardioverter-defibrillator (ICD) a surgically implanted device that can detect and cardiovert an abnormal cardiac rhythm, such as ventricular tachycardia or ventricular fibrillation.

Implied consent see **best interests standard of consent**.

In Vitro Fertilization (IVF) a reproductive technology to treat infertility, where sperm and ovum are combined in a laboratory. The resulting embryos are subsequently implanted into a woman's uterus.

Infanticide the deliberate killing of a human baby.

Infertility the inability to achieve a pregnancy after attempting for one year or more.

Informed consent the three-element standard to allow a clinician to proceed with a medical treatment or procedure: full disclosure, patient comprehension, and documented permission.

Institutional Review Board (IRB) a deliberative committee at a clinical research institution that considers the ethics of specific proposed research projects involving human subjects.

Intact dilation and extraction an abortion procedure that employs a feet-first delivery, followed by puncture decompression of the skull to permit easier removal of the now-dead fetus. Also known by the non-technical phrase "partial-birth abortion," this procedure is illegal in the U.S.

Interceptive effect a possible but unproven mechanism of some contraceptives, where the drug interferes with the implantation of an early human embryo.

Intracytoplasmic Sperm Injection (ICSI) an assisted reproductive procedure that uses a single sperm cell injected directly into the cytoplasm of an oocyte to achieve fertilization.

Intrauterine Insemination (IUI) also known by the more general term **artificial insemination**, an assisted reproductive procedure that instills concentrated sperm directly into a woman's uterus.

Kantian ethics a reason-based ethical framework first proposed by Immanuel Kant (see **categorical imperative**).

Lebensunwertes leben German phrase for "life unworthy of life," used by Karl Binding and Alfred Hoche in 1920 to justify the medical killing of disabled persons.

Levonorgestrel a commonly-used progestin in combined oral contraceptives, and the single agent used in the most common form of emergency contraception (available without a prescription).

Living will an advance directive that lists specific medical treatment preferences in the event of patient incapacitation.

Mature minor doctrine the rule that an adolescent minor 15 years of age or older who appears to make reasoned judgments may consent to or refuse a proposed medical treatment for the patient's benefit, when parental consent cannot be obtained.

Medical aid in dying see **assisted suicide**.

Medical futility the determination that further aggressive medical therapy will not prolong life or ease suffering, an easily abused concept.

Medical Orders for Life-Sustaining Treatments (MOLST) see **Physician Orders for Life-Sustaining Treatments**.

Medical principlism a widely-recognized framework for medical ethics, defined by four rules: beneficence, non-maleficence, distributive justice, and autonomy.

Medical quackery a form of treatment that has no potential for benefit, the known harm of a treatment is significantly greater than any potential benefit, or those promoting the treatment are being deliberately deceptive, usually for profit.

Medication therapy management (MTM) a pharmacist review of all prescribed and over-the-counter medications in a given patient, in order to identify and address medication problems.

Meta-ethics the study of how moral philosophy is defined, examining such moral concepts as right, wrong, and justice.

Mifepristone a progesterone-receptor antagonist, given as the first of two medications to induce a medical abortion.

Minderwertig a German term used in the years before and during World War II to identify genetically "inferior" individuals.

Misoprostol a prostaglandin that induces uterine contractions, used as the second drug to accomplish a medical abortion. It is also often used to deliver uterine contents following a spontaneous miscarriage.

mitochondrial DNA (mtDNA) in contrast to nuclear DNA, mtDNA is found out in the cytoplasm, within the small circular chromosomes of mitochondria.

Mitochondrial transfer technology an assisted reproductive technology that utilizes mitochondria from donor oocytes to replace a woman's abnormal mitochondria, preventing a heritable mitochondrial disease in the offspring.

Moral agent see **person**.

Moral complicity the possible taint of moral guilt attached to a person by direct or indirect association with a moral wrong.

Moral distress for a healthcare professional, a disruptive emotional state due to an unresolvable moral dilemma, particularly where the clinician knows the morally right thing to do but cannot do it because of legal or situational barriers.

Moral taint see **moral complicity**.

Morality a set of principles of right and wrong behavior, often used as a synonym for ethics.

Morally heroic a voluntary, beneficial act that goes beyond mere duty.

Morally neutral a voluntary act with no moral significance.

Morally obligatory a voluntary, beneficial act that is ethically required.

Morning-after pill see **emergency contraception**.

Natural family planning non-hormonal family planning that relies on the timing of sexual intimacy to either avoid or promote pregnancy.

Natural law ethics the reason-based theory that morality is built into the nature of human beings and is intuitively knowable.

Non-maleficence the avoidance of harm.

Normative ethics a branch of ethics whose standards consist of widely-shared norms about right and wrong human conduct and form a stable social consensus.

Nurturing mother the woman who feeds and otherwise cares for a child, even if the child is not hers biologically or gestationally.

Oligospermia an abnormally low sperm count, which may lead to infertility.

Ontological personalism the philosophical idea that all human beings are human persons, by virtue of membership in the human species.

Palliative care treatment oriented primarily towards the relief of symptoms rather than cure.

Pandemic a breakout of an infectious disease that is spread out over many countries.

Partial-birth abortion see **intact dilation and extraction**.

Paternalism an authoritarian style of medical care that undermines individual autonomy.

Persistent Vegetative State (PVS) a prolonged state of unresponsiveness due to cerebral cortical damage with an intact and functioning brainstem.

Person a member of the moral community, a term that indicates individual moral value.

Pharmacogenomics the study of how an individual's genetic makeup influences his or her response to certain medications.

Pharmacy benefit manager (PBM) a third-party administrator of a prescription drug program that specifically manages the pharmaceutical benefit on behalf of the plan sponsor.

Physician Orders for Life-Sustaining Treatments (POLST) a type of advance directive wherein patients with decisional capacity and a life expectancy of less than a year may outline their treatment preferences. POLST forms become an actionable set of medical orders that transfer across institutions and include direct orders for paramedics and nurses.

Placebo a functionally inert and non-therapeutic pill or injection, often used in randomized controlled clinical research trials, to determine the therapeutic efficacy of a particular medication.

Planned Parenthood of Southeastern Pennsylvania v. Casey the 1992 decision by the U.S. Supreme Court that discarded the trimester system established in *Roe v. Wade*, but also threw out a spousal notification requirement, declaring that it imposed an undue burden on women seeking an abortion.

Precision medicine also called personalized medicine, primarily using genetic data and molecular approaches to target specific therapies in patients to maximize benefit and reduce risk (see **pharmacogenomics**).

Preimplantation Genetic Diagnosis (PGD) also referred to as embryo biopsy, the genetic analysis of a three-day human embryo to select out undesirable genetic defects.

Prima facie argument from the Latin for "first face," refers to an immediate or first impression ethical assumption in evaluating the ethics of a clinical situation or issue.

Principle of double effect (see **double effect**).

Privacy right though not found directly in the U.S. Constitution, a right implied by specific guarantees in the Bill of Rights (first articulated in *Griswold v. Connecticut*). This right would later form an important basis for the majority opinion in *Roe v. Wade*.

Property-thing an entity defined by a collection of parts and functions, one possible understanding of human nature.

Proportionality an ethical framework for evaluating a proposed clinical treatment plan by comparing the benefits of the treatment to realizable goals.

Quality of life the perceived value of continued existence, as assessed by an individual or family members.

Rassenhygiene (racial hygiene) a vague concept from Nazi ideology that focused on "cleansing" the German people through various eugenics measures.

Reproductive tourism the practice of traveling to another country to obtain assisted reproductive technologies or surrogacy arrangements (cp. **circumvention medical tourism**).

Respondeat superior applied from civil law specifically to pharmacy practice, this phrase means that the fiduciary duty of pharmacists to their patients cannot be delegated to technical or clerical personnel.

right of conscience a right of a clinician to not participate in a particular treatment plan that violates the clinician's personal moral or religious values.

Roe v. Wade the landmark 1973 U.S. Supreme Court decision that changed abortion-related statutes in 46 states and made elective abortion legal within the first two trimesters of pregnancy.

Savior sibling the idea of using reproductive technologies and preimplantation genetic diagnosis to select and gestate a new baby that could serve as a tissue donor to treat an existing child with a genetic disease.

Secobarbital a barbiturate sedative commonly used for physician-assisted suicide.

Situation ethics a relationship-based ethical theory popular in the 1970s, first proposed by Joseph Fletcher. Situation ethics considers the circumstances of a particular action to maximize love.

Slippery-slope argument a claim that the acceptance of a specific controversial practice will open the door to greater and greater deviations from what is morally acceptable. Such an argument without substantial evidence to back it up can be a logical fallacy.

Social Darwinism the application of the scientific theory of Darwinian evolution to the social sciences. This idea was a significant influence on the eugenics movement.

Somatic cell gene therapy the modification of the DNA of a child or adult, the effects of which are confined to that individual and cannot be passed on to future generations.

Speciesism a term coined by Professor Peter Singer of Princeton University to refer to bias or prejudice against non-human animals and rejecting them as having moral value.

Strict liability in civil law, the legal standard that applies to a clinician concerning errors for which he or she alone is responsible.

Substance from philosophy, a distinct unity of essence that exists ontologically prior to any of its parts. This concept is used in arguments for the personhood of all human beings at every stage of their development.

Substituted judgment a decision made by a surrogate (**healthcare proxy**) on behalf of an incapacitated patient.

Suction aspiration an abortion technique that uses a manual or machine vacuum curette to remove the fetus. It is also used to deliver uterine contents following a spontaneous miscarriage.

Surrogate an alternative decision-maker for an incapacitated patient (sometimes called a **healthcare proxy**, see **durable power of attorney for health care**).

Surrogate motherhood a formal or informal arrangement wherein a woman agrees to gestate a child on behalf of another couple (see also **commercial surrogacy** and **altruistic surrogacy**).

Tech-check-tech a procedure where a pharmacy technician verifies the preparation of medications by another pharmacy technician.

Telemedicine the practice of providing remote access to healthcare via Internet technology.

Terminal condition a general term for a disease or process that will eventually cause the premature death of a person.

Theological voluntarism see **divine command theory**.

Thimerosal a compound containing organic mercury, used as a preservative in some vaccines.

Three-person embryos see **mitochondrial transfer technology**.

Triage in the context of a war or mass casualty event, the principle of establishing patient treatment priority. The goal is to save the most lives with limited resources.

Trolley Problem a classical thought experiment in ethics that uses the example of an out of control railway trolley to highlight deontological versus consequentialist thinking.

U.S. Pharmacopeia first established in 1820, the annual compendium of drug information in the United States.

Ulipristal a prescription-only emergency contraceptive agent that acts as a potent modulator of progesterone receptors, leading to the reversible inhibition of progestin.

Utilitarianism in its classical form, the ethical theory that claims the right action as that which produces the most happiness for the greatest number of people.

Vaccination injecting biological material to induce immunity against a specific infectious disease.

Variolation a now obsolete practice of producing immunity to smallpox by injecting someone with infected material from someone with a mild case.

Virtue ethics an ethical theory first proposed by Aristotle that focuses on character rather than on the solving of particular ethical problems.

Withdrawing treatment the removal of an unwanted or ineffective clinical treatment.

Withholding treatment the decision not to use a specific treatment, usually because of a perception that it would not be effective or that the harms would outweigh the benefits.

Index

A
Abortifacient, 89, 146
Abortion, 91–94
Abstinence, 88
Active euthanasia, 113
Advance directives, 44, 118–121
Aesculapius, 37, 142
After-birth abortion, 29
Altruistic surrogacy, 102
American Eugenics Society, 51
Apollo, 142
Apothecaries, 5
Aquinas, T., 15, 88
Arendt, H., 58
Aristotle, 22
Artificial insemination, 95
Artificial intelligence (AI), 214–215
Assent, 45, 46
Assisted reproductive technologies (ART), 95–100
Assisted suicide, 125, 126
Autism, 152
Autonomy, 37, 42, 66, 70, 75, 77, 113, 125, 129, 132–134, 138, 144, 159, 164, 177, 180, 182, 183

B
Baby M, 100
Banality of evil, 58
Belmont Report, 69, 70
Beneficence, 2, 37, 39, 66, 70, 77, 113, 129, 142, 144, 159, 164, 177, 182, 193
Bentham, J., 20

Best interests, 44
Brain death, 114, 115, 189
BRCA, 211
Brown, L.J., 95
Buck v. Bell, 52, 53

C
Carrie Buck, 52
Categorical imperative, 14, 59, 128
Certiorari, 200
Cicely Saunders, 110
Circumvention medical tourism, 210
Clinical equipoise, 70, 162, 213
Code of Ethics for Pharmacists, 75
Collaborative practice agreements (CPAs), 82, 191
Collins, F.S., 205
Combined oral contraceptive (COC), 89
Comfort care, 109
Commercial surrogacy, 102
Competency, 43
Competing duties, 3
Confidentiality, 41
Conscientious objection, 141
Consequentialist, 12
Contraception, 87–91
Coronavirus disease 2019 (COVID-19), 158, 162, 164
Corresponding responsibility, 83
Covenant, 76, 79
Covenantal, 137
Crisis standards of care, 159, 169
CRISPR, 206

D

Darwin, C., 50
Dead Donor Rule, 114, 115
Decision-making capacity, 42–44, 119, 189
Declaration of Independence, 15
Deontological, 12, 128, 177
Designer babies, 99
Dilation and evacuation, 94
Distributive justice, 37, 41, 66, 70, 78, 113, 129, 144, 159, 164, 177
Divine command theory, 16, 17, 177, 186, 199
Doctors' Trial, 58, 68
Down syndrome, 98
Drug Importation Act, 5
Durable power of attorney for healthcare (DPAHC), 44, 118, 120

E

Ectopic pregnancy, 94
Edward Jenner, 151
Eighth Amendment, 2, 136–138, 198
Emancipated minor, 46
Embryo donation, 97
Emergency contraception (EC), 89
Emergency Use Authorization (EUA), 166
Empirical functionalism, 27, 58
Enlightenment, 41
Ethics, 9
Eugenics, 49, 63
Eugenics Record Office (ERO), 51
Euthanasia, 55, 125
Euthyphro, 18

F

Federal Wide Assurance, 71
Feeblemindedness, 50, 52
Fetal tissue, 154, 168
Fiduciary, 76, 79, 83, 137, 142, 190
Final Solution, 55
Fletcher, J., 22, 27
Food, Drug, and Cosmetic Act, 5, 137, 163
Fourteenth Amendment, 93
Freedom of action, 10
Frozen embryos, 97
Futility, 112, 195

G

Galton, F., 49
Gene editing, 206–208
Genetic Information Non-Discrimination Act (GINA), 211
Genetic testing, 211
Germline treatments, 208
Gonzales v. Carhart, 93
Griswold v. Connecticut, 92

H

Health Insurance Portability and Accountability Act (HIPAA), 71, 147, 212
Healthcare proxy, 120
Helsinki Declaration, 68
Hepatitis A, 67
Herd immunity, 156, 167
Hippocrates, 2, 19, 37
Hippocratic Oath, 37, 69, 75, 76, 132, 141
Holmes, O.W., 53
Holocaust, 56
Hospice care, 111, 132, 134, 197
Human dignity, 32, 33, 130, 132
Human Genome Project, 205, 206
Human non-person, 28
Hypothetical imperative, 14

I

Imminent death, 112
Implied consent, 44
Infertility, 95
Informed consent, 42–44, 159, 183
Institutional Review Board (IRB), 69–72
Intact dilation and extraction, 93
Interceptive, 89
Intracytoplasmic sperm injection (ICSI), 96
Intrauterine insemination (IUI), 95
In vitro fertilization (IVF), 95, 209

J

Jefferson, Thomas, 15
Jehovah's Witness, 46, 177
Jurisprudence, 5

K

Kantian ethics, 13, 99, 177, 199
Kant, Immanuel, 14, 41, 128
Kűbler-Ross, E., 110

L

Lebensunwertes leben, 54, 58
Lethal injection, 3, 135–138, 198
Levonorgestrel, 90

Index 233

Liability, 193
Living will, 118
Locke, J., 15

M
Mature minor, 46
Medical principlism, 19, 37, 69, 128, 138, 144, 177
Medication therapy management (MTM), 80
Mendel, Gregor, 50
Messenger RNA, 166
Meta-ethics, 9
Mifepristone, 94
Mill, J.S., 20
Minderwertig, 55
Misoprostol, 94, 146
Mitochondrial DNA (mtDNA), 208
Mitochondrial transfer technology, 209
MMR vaccine, 153
Moral agent, 27
Moral complicity, 147, 154, 199
Moral distress, 188
Morality, 9
Moral philosophy, 9

N
Nancy Cruzan, 116
Natural law, 15, 88, 118
Negative eugenics, 50, 99
Non-maleficence, 2, 37, 39, 66, 77, 113, 128, 138, 142, 144, 159, 164, 177, 193
Normative ethics, 9, 10
Nuremberg code, 68

O
Oath and Prayer of Maimonides, 19, 128
Office of Human Research Protections (OHRP), 71
Ontological personalism, 27, 29, 58
Operation Warp Speed (OWS), 166
Opioid addiction, 187, 197, 216

P
Palliative care, 110, 132, 134, 184
Palliative sedation, 184
Pandemic, 158–161, 164–169, 192
Partial-birth abortion, 93, 94
Pediatric consent, 45, 46
Persistent vegetative state (PVS), 116, 189
Personal dignity, 130, 132

Personhood, 27
Pharmacogenomics, 212
Pharmacy Benefit Managers (PBMs), 215
Pharmacy Code of Ethics, 198
Physician-assisted suicide (PAS), 125, 126
Physician Orders for Life-Sustaining Treatments, 120
Placebos, 45
Planned Parenthood of Southeastern Pennsylvania v. Casey, 93
Plato, 18, 49
Positive eugenics, 50, 99
Precision medicine, 5, 212, 213
Preimplantation genetic diagnosis (PGD), 98
Prima facie, 31
Primum non nocere, 39
Principle of double effect, 113
Profession, 148
Professional ethics, 9, 10
Professionalism, 81–83, 141, 143, 164
Property-thing, 30
Proportionality, 118
Provider status, 5
Pure Food and Drug Act, 5, 163
Pythagoras, 37

Q
Quackery, 162
Quality of life, 112, 177
Quinlan, Karen Ann, 116

R
Rassenhygiene, 54, 58
Reason-based systems, 13
Relationship-based systems, 13
Reproductive tourism, 100, 210
Respondeat superior, 83
Revised Common Rule, 71
Rhythm method, 88
Rights of conscience, 141, 144, 200
Roe v. Wade, 92

S
Singer, P., 28, 29, 95, 133
Situation ethics, 22
Slippery slope, 129, 131, 132, 134
Social Darwinism, 50
Social distancing, 166
Somatic cell gene therapy, 207
Speciesism, 29
Stem-cell research, 98

Sterilization laws, 52
Strict liability, 83
Substance, 30
Substituted judgment, 44
Suction aspiration, 94
Sulmasy, D.P., 130
Surrogacy, 44, 100–102, 183, 191
Syphilis, 64

T
Telemedicine, 215
Terminal condition, 112
Terri Schiavo, 117
Theological voluntarism, 17
Thimerosal, 153
Three-Person Embryos, 208–210
Tooley, Michael, 28
Tradition-based systems, 13
Triage, 159, 169
Trolley problem, 11
Tuskegee Syphilis Study, 63–67

U
Ulipristal, 90
Unintended consequence, 113
Universal Declaration of Human Rights, 32
Utilitarianism, 20, 160, 177, 199

V
Vaccines, 151–158
Ventilator allocation guidelines, 160
Virginia Colony, 52
Virtue ethics, 22, 79, 80, 144, 177, 199

W
Wakefield, A.J., 152
Wellness promotion, 80
Wiesel, E., 56, 58
William of Ockham, 18
Willowbrook Hepatitis Studies, 67, 68
Withdrawing treatment, 113
Withholding treatment, 113

The manufacturer's authorised representative in the EU is Springer Nature Customer Service Centre GmbH, Europaplatz 3, 69115 Heidelberg, Germany. If you have any concerns regarding our products, please contact ProductSafety@springernature.com

Printed and bound by CPI Group (UK) Ltd, Croydon, CR0 4YY

25/03/2026

02078169-0002